Analysis and Design of Resilient VLSI Circuits

Rajesh Garg · Sunil P. Khatri

Analysis and Design of Resilient VLSI Circuits

Mitigating Soft Errors and Process Variations

Springer

Rajesh Garg
6430 NE Alder St.
Apt. B
Hillsboro, OR 97124
USA
rajeshgarg@tamu.edu

Sunil P. Khatri
Department of Electrical and Computer Engineering
Texas A & M University
214 Zachry Engineering Center
College Station, TX 77843-3128
USA
sunilkhatri@tamu.edu

ISBN 978-1-4419-0930-5 e-ISBN 978-1-4419-0931-2
DOI 10.1007/978-1-4419-0931-2
Springer New York Dordrecht Heidelberg London

Library of Congress Control Number: 2009936000

Printed on acid-free paper

Springer is part of Springer Science+Business Media (www.springer.com)

To,

Our families

-Rajesh and Sunil

Preface

This monograph is motivated by the challenges faced in designing reliable VLSI systems in modern VLSI processes. The reliable operation of integrated circuits (ICs) has become increasingly difficult to achieve in the deep submicron (DSM) era. With continuously decreasing device feature sizes, combined with lower supply voltages and higher operating frequencies, the noise immunity of VLSI circuits is decreasing alarmingly. Thus, VLSI circuits are becoming more vulnerable to noise effects such as crosstalk, power supply variations, and radiation-induced soft errors. Among these noise sources, soft errors (or error caused by radiation particle strikes) have become an increasingly troublesome issue for memory arrays as well as combinational logic circuits. Also, in the DSM era, process variations are increasing at a significant rate, making it more difficult to design reliable VLSI circuits. Hence, it is important to efficiently design robust VLSI circuits that are resilient to radiation particle strikes and process variations. The work presented in this research monograph presents several analysis and design techniques with the goal of realizing VLSI circuits, which are radiation and process variation tolerant.

This monograph consists of two parts. The first part proposes four analysis and two design approaches to address radiation particle strikes. The analysis techniques for the radiation particle strikes include: an approach to analytically determine the pulse width and the pulse shape of a radiation-induced voltage glitch in combinational circuits, a technique to model the dynamic stability of SRAMs, and a 3D device-level analysis of the radiation tolerance of voltage scaled circuits. Experimental results demonstrate that the proposed techniques for analyzing the effect of radiation particle strikes in combinational circuits and SRAMs are fast and accurate when compared with SPICE simulations. Therefore, these analysis approaches can be easily integrated in a VLSI design flow to analyze the radiation tolerance of ICs, to harden them early in the design flow. From 3D device-level analysis of the radiation tolerance of voltage scaled circuits, several nonintuitive observations are made and correspondingly, a set of guidelines are proposed, which are important to consider in order to realize radiation hardened circuits. In the first part of this monograph, two circuit level hardening approaches are also presented to harden combinational circuits against a radiation particle strike. These hardening approaches significantly improve the tolerance of combinational circuits against low and very high energy radiation particle strikes, respectively, with modest area and delay overheads.

The second part of this monograph addresses process variations. A technique is developed to perform sensitizable statistical timing analysis of a circuit, and thereby it improves the accuracy of timing analysis under process variations. Experimental results demonstrate that this technique is able to significantly reduce the pessimism due to two sources of inaccuracy, which plague current statistical static timing analysis (SSTA) tools. Two design approaches are also proposed to improve the process variation tolerance of combinational circuits and voltage level shifters (which are required in circuits with multiple interacting power supply domains), respectively. The variation tolerant design approach for combinational circuits significantly improves the resilience of these circuits to random process variations, with a reduction in the worst case delay and with a low area penalty. The proposed voltage level shifter is faster, requires lower dynamic power and area, has lower leakage currents, and is more tolerant to process variations, compared with the best known previous approach.

In summary, this monograph presents several analysis and design techniques which significantly augment the existing body of knowledge in the area of resilient VLSI circuit design.

Hillsboro, OR — *Rajesh Garg*
College Station, TX — *Sunil P. Khatri*
April 2009

Acknowledgements

The authors would like to gratefully acknowledge Dr. Nikhil Jayakumar, Kanupriya Gulati, Charu Nagpal, and Gagandeep Mallarapu for their direct or indirect contribution to the work presented in this monograph. The efforts of Dr. Nikhil Jayakumar, who worked jointly on the diode clamping-based radiation hardening approach and the sensitizable statistical timing analysis approach, are kindly appreciated. Work on the analytical determination of radiation-induced pulse width and on the voltage level shifter was performed jointly with Charu and Gagan, respectively. The insightful comments of Dr. Peng Li, Dr. Hank Walker, Dr. Kevin Nowka, Dr. Gwan Choi, and Dr. Krishna Narayanan are gratefully acknowledged.

Contents

List of Figures

List of Tables

Acronyms

CVLS	Conventional voltage level shifter
DRAM	Dynamic random access memory
LET	Linear energy transfer
PVT	Process voltage and temperature
SEE	Single event effect
SEU	Single event upset
SET	Single event transient
SFM	Strong feedback mode
SRAM	Static random access memory
STA	Static timing analysis
SSTA	Statistical static timing analysis
SS-VLS	Single supply voltage level shifter
SS-TVLS	Single supply true voltage level shifter
StatSense	Sensitizable statistical timing analysis
WCM	Weak coupling mode
GND	Ground
τ_α	Charge collection time constant
τ_β	Ion track establishment constant
C_G	Gate capacitance
C_D	Output node diffusion capacitance
H	Wire height
$i_{seu}(t)$	Radiation-induced current pulse
I_{DS}	Drain to source current
L	Channel length
Q_D	Charge deposited by a radiation particle
Q	Charge collected at a node due to a radiation particle strike
P_{LM}	Probability of logical masking
P_{Sen}^G	Probability of sensitization of gate G
R_n	Linear resistance of NMOS transistor
V_{diode}	Diode turn on voltage
V_{dsat}^N	Saturation voltage of NMOS transistor
V_{dsat}^P	Saturation voltage of PMOS transistor
V_{GM}	Maximum voltage glitch magnitude

T_{ox}	Oxide thickness
V_{ST}	Gate switching threshold voltage
V_T	Threshold voltage
V_{TP}	Threshold voltage of PMOS transistor
V_{TN}	Threshold voltage of NMOS transistor
W_M	Wire width
VDD	Supply voltage

Chapter 1
Introduction

Reliability of very large scale integration (VLSI) systems has always been a major concern. Integrated circuits (ICs) have always been subjected to several reliability degrading factors such as manufacturing defects (e.g., wire shorts, wire opens, etc.), electromigration, noise, etc. To deal with these issues, various forms of fault tolerance have been built into digital systems for the past several decades. Recently, in the deep sub-micron (DSM) era, with continuously decreasing device feature sizes, lowering supply voltages, and increasing operating frequencies, the tolerance of VLSI systems against these effects has significantly decreased. In addition to this, several new factors such as process variations, aging, etc. now further adversely affect digital VLSI system reliability. Therefore, in the DSM regime, the design of reliable digital VLSI systems has become very challenging.

There are many types of noise effects in VLSI systems, such as power and ground noise, capacitive coupling noise, radiation particle strikes or single event effects (SEEs), etc. With technology scaling, ICs have become very sensitive to radiation particle strikes [1, 2, 3, 4, 5, 6, 7]. Radiation particle strikes affect the transient electrical behavior of a circuit and can result in functional errors. Such errors are often referred to as *soft* or *transient errors*. Researchers expect an approximate 8% increase in soft error rate (SER) per logic state bit for each technology generation [6, 7]. Also, the number of logic state bits on a chip doubles each technology generation. This further increases the sensitivity of ICs to radiation particle strikes with technology scaling. It is expected that the SER for chips implemented in the 16 nm technology will be almost 100× of the SER of chips implemented in the 180 nm technology [6, 7]. Also, with device scaling, the variations of key device parameters are increasing at an alarming rate [8, 9, 10], making it difficult to predict the performance of a VLSI design. Thus, both these issues (radiation particle strikes and process variations) result in unpredictable behavior of circuits and hence severely degrade the reliability of VLSI systems. Because of the widespread use of modern VLSI systems, it is necessary to address these issues during the design phase, to improve system reliability and resilience to radiations and process variations. This is the focus of this monograph.

In the remainder of this chapter, Sect. 1.1 provides background information about radiation particle strikes and process variations. It also describes how these issues affect VLSI circuit operation. The goals of the research work presented in this

R. Garg and S. P. Khatri, *Analysis and Design of Resilient VLSI Circuits: Mitigating Soft Errors and Process Variations*, DOI 10.1007/978-1-4419-0931-2_1,

monograph are stated in Sect. 1.2. Section 1.2 also provides an outline of the remaining chapters of this monograph. Finally, a chapter summary is provided in Sect. 1.3.

1.1 Background and Motivation

This section provides some background information about radiation particle strikes and process variations, to aid in understanding the remainder of this monograph. It also describes how these issues affect VLSI system operation, and how they are expected to scale in future technologies.

1.1.1 Radiation Particle Strikes

SEEs are caused when radiation particles such as protons, neutrons, alpha particles, or heavy ions strike sensitive regions (usually reverse-biased p-n junctions) in VLSI designs. These radiation particles strikes can deposit a charge, resulting in a voltage pulse or glitch at the affected node. This radiation-induced voltage glitch can result in a soft or transient error.

Radiation particle strikes are very problematic for memories (latches, SRAMs, and DRAMs) since they can directly flip the stored state of a memory element, resulting in a *single event upset (SEU)* [1, 2]. Although radiation-induced errors in sequential elements will continue to be problematic for high performance microprocessors, it is expected that soft errors in combinational logic will dominate in future technologies [4, 11, 12], as discussed later. Radiation strikes in combinational circuits are referred to as *single event transients (SETs)*. In a combinational circuit, a voltage glitch due to a radiation particle strike can propagate to the primary output(s) of the circuit, which can result in an incorrect value being latched by the sequential element(s), hence resulting in single or multiple bit upsets. Whether or not a voltage glitch induced by a radiation particle strike at any gate in a combinational circuit propagates to the primary outputs (and results in a failure) depends upon three masking factors. These masking factors are [4, 12]:

- *Electrical masking* occurs when a voltage glitch at a circuit node induced by a radiation particle strike attenuates as it propagates through the circuit to the primary outputs. Electrical masking can reduce the voltage glitch magnitude to a value which cannot cause any soft errors.
- *Logical masking* occurs when there is no functionally sensitizable path from the node in the circuit where a radiation particle strikes, to any primary output of the circuit. Hence, logical masking properties of a gate can be estimated using logic information alone.
- *Temporal masking* occurs if a voltage glitch due to a radiation particle strike reaches the primary outputs of a circuit at an instant other than the latching window of the sequential elements of the circuit. Temporal masking only depends

upon the frequency of operation of the circuit. Its influence is identical for all gates in the circuit (for a given voltage glitch due to a particle strike). Therefore, it provides the circuit some gratuitous radiation tolerance against soft errors.

Note that all these masking factors reduce the severity of a radiation particle strike in combinational circuits. In other words, if a gate in a circuit is masked to a large extent by any of these masking factors, then it is unlikely (low probability) that a radiation particle strike at the output of that gate will have any effect on the primary outputs of the circuit. Only those gates in a combinational circuit that exhibit a low degree of masking due to these three factors (referred to as *sensitive gates*) contribute significantly to the failure of the circuit due to soft errors.

Until recently, radiation particle strikes were considered troublesome only for military and space electronics. This is mainly due to the abundance of radiation particles in the operating environment of such systems. In fact, the first confirmed radiation-induced upsets in space (four upsets in 17 years of satellite operation) was reported in 1975 [13]. However, just 4 years later (i.e., in 1979), soft errors were also observed in terrestrial microelectronics [1]. Since then, with technology scaling, several cases of soft errors or upsets have been observed in both space as well as terrestrial electronics [11]. Therefore, for applications such as space, military, and critical terrestrial (for example biomedical) electronics, which place a stringent demand on reliable circuit operation, it is important to use radiation tolerant circuits. To efficiently design radiation tolerant circuits, it is important to understand the effects of radiation particle strikes on VLSI systems.

The rest of this section is devoted to a discussion on the physical origin of radiation particles, how these particle strikes result in voltage transients, the modeling of a radiation particle strike in circuit level simulations, and the impact of technology scaling on the sensitivity of VLSI designs to radiation particle strikes.

1.1.1.1 Physical Origin of Radiation Particles

In space, the cosmic rays enter the solar system from the outside which are referred to as galatic cosmic rays. These rays are high-energy charged particles composed of protons, electrons, and heavier nuclei [14]. These energy particles are primarily responsible for soft errors in space electronics [11]. Apart from galactic cosmic rays, solar event protons and protons trapped in the earth's radiation belts are the other sources of protons present in the earth's atmosphere [11]. These are also capable of producing SEEs. Alpha particles may also originate from radioactive contaminations in IC packages [11]. In fact, the first soft error reported for terrestrial electronics [1] was due to alpha-particles that originated from IC packaging materials. Recently, flip-chip packages have been identified as a source of radiation particles (from the Pb-Sn solder bumps). This aggravates the problem of radiation hardening because a source of radiation particles is present extremely close to the die. Also at the surface of the earth, neutrons induced upsets have found to be very problematic. Several studies have found that the neutrons from cosmic rays are a significant source of soft errors for SRAMs and DRAMs [11] operating at the earth's surface. These atmospheric neutrons result when high energy galactic cosmic rays collide with other particles in the earth's atmosphere. Thus, the neutron flux varies a lot with altitude and latitude [11,4,15]. The authors of [4] reported that the neutron flux at an altitude of 10,000 feet in Leadville, CO is approximately 13× greater than that at the sea level. Because of this, a large number of neutron induced upsets were observed in DRAMs at 10,000 feet in Leadville, CO, while no upsets were observed when the DRAM was placed 200 m underground in a salt mine [11].

Different radiation particles such as protons, neutrons, alpha-particles, and heavy ions have different mechanisms by which they deposit charge in VLSI designs. These mechanisms are explained next.

1.1.1.2 Charge Deposition Mechanisms

There are two methods by which a radiation particle deposits charge in VLSI designs: direct ionization and indirect ionization.

Direct Ionization. A radiation particle generates electron–hole pairs along its path as it passes through a semiconductor material, as shown in Fig. 1.1. In this process, the radiation particle loses its energy. After losing all its energy, the particle comes to rest. The energy transferred by the radiation particle is described by its linear energy transfer (LET) value. LET is defined as the energy transferred (for electron–hole pair generation) by the radiation particle per unit length, normalized by the density of the target material (for VLSI designs, this is the density of Silicon). Thus the unit of LET is MeV-cm^2/mg. The LET of a radiation particle also corresponds to the charge deposited by the radiation particle per unit length. In silicon, the amount of charge deposited (Q_{D}) by a radiation particle per unit length (in microns) is calculated as $Q_{\mathrm{D}} = 0.01036 \cdot LET$. For example, a particle with an LET of 97 MeV-cm^2/mg

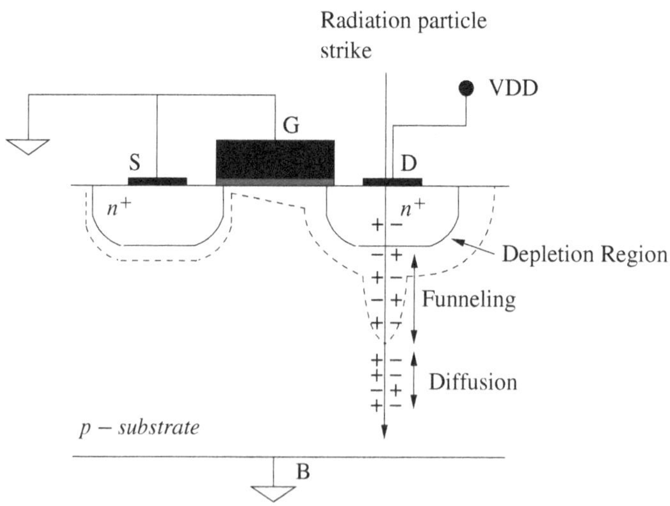

Fig. 1.1 Charge deposition and collection by a radiation particle strike

can deposit 1 pC/μm. Heavy ions[1] and alpha-particles primarily deposit charge in a semiconductor by direct ionization. Light particles such as protons and neutrons do not deposit enough charge by direct ionization to cause a soft error.

Indirect Ionization. Protons and neutrons typically deposit charge by indirect ionization, which can result in significant numbers of soft errors [11, 4, 16]. When a high-energy light radiation particle (such as a proton or a neutron) passes through a semiconductor material, it can collide with nuclei, resulting in nuclear reactions. These nuclear reactions may produce secondary particles such as alpha-particles or heavy ions. These secondary particles then deposit charge by direction ionization and if the charge is deposited at different locations in a chip then multiple soft errors may occur [11, 16]. Thus, the charge deposited by a light particle through indirect ionization heavily depends upon the location and the angle of incidence of the particle strike.

When charge is deposited due to a radiation event, this charge is *collected* by different terminals of the devices, resulting in voltage and current transients in the device. The charge deposited by a radiation particle strike may get collected through different charge collection mechanisms which are briefly described next.

1.1.1.3 Charge Collection Mechanisms

There are three charge collection mechanisms as discussed below:

Drift-diffusion. Consider an NMOS transistor shown in Fig. 1.1. The source, gate, and bulk terminals of the NMOS transistor are connected to GND. The drain terminal is connected to VDD. The drain-bulk junction is reverse-biased and hence there is a strong electric field in the depletion region of this junction from the drain to the bulk. Since radiation particle generated free electron–hole pairs, the electric field present in the depletion region of the drain-bulk junction leads to the collection of electrons at the drain and of holes at the bulk. Thus, the reverse-biased electric field leads to the charge collection at the drain. Therefore, the reverse biased junctions are most sensitive to a radiation particle strike. Assume that a radiation particle strikes this (drain-bulk) junction and generates electron–hole pairs along its path as shown in Fig. 1.1. Immediately after the generation of this ionized track, the depletion region collapses due to the separation of free electrons and holes by the drift process in the depletion region. As mentioned earlier, charge (electrons and holes) separation occurs due to the presence of a high electric field, which pulls the electrons up (toward the n^+ diffusion) and pushes the holes down (toward the p-substrate). This phenomenon reduces the width of the depletion region of the drain-bulk junction. As a result, the potential drop across the depletion region decreases (before the radiation strike, the potential drop across the depletion region was VDD). As the voltage between the drain and the bulk terminals (n^+ and p-substrate) is still VDD, the decrease in the potential across the depletion

[1] Heavy ion are ions whose atomic number is greater than or equal to 2 [11].

region causes a voltage drop in the p-substrate region. This causes the drain-bulk junction electric field to penetrate into the p-substrate region, beyond the original depletion region and hence enhances the flow of electrons from the substrate (these electrons are generated by the radiation particle strike in the substrate region) to the depletion region. This enhanced electron flow process is referred to as *funneling* as shown in Fig. 1.1. The electrons present in the depletion region drift to the drain (n^+) diffusion region and hence get collected. Thus, charge is said to be collected through the *drift* process (or the *funnel-assisted drift process*). The funneling process increases the depth of the region with a strong electric field beyond the original depletion region. Hence it increases the amount of charge collection by the drift process [17, 18, 19, 11].

As the electric field continues to pull electrons up, it also pushes the holes down (away from the depletion region) which allows the drain-bulk depletion region to recover and regain its original width. After the recovery of the depletion region, the electrons that were not collected by the funnel-assisted drift process diffuse toward the depletion region (due to their concentration gradient) and then get pulled by the junction electric field toward the n^+ drain diffusion region. Thus the charge is also collected at the n^+ region by the *diffusion* process. It was reported in [18], that in a lightly-doped substrate, most of the charge collection is through drift only, whereas in more heavily-doped substrates demonstrate charge collection due to both the drift and the diffusion processes [20, 17, 18, 19, 11]. In the DSM technologies, the substrate is heavily doped and hence charge collection at the drain node occurs due to both drift and diffusion processes.

Bipolar Effect. Consider an NMOS transistor (an n-channel transistor located in a p-well) in cut-off state, and with its gate and source terminals at GND and drain terminal at VDD. The electrons generated by a radiation particle strike can be collected at either the drain-well junction or the well-substrate junction. However, the radiation-induced holes are left in the p-well, which reduces the source-well potential barrier (due to the increase in the potential of the p-well). Thus, the source injects electrons into the channel which can be collected at the drain. This increases the total amount of the charge collected at the drain node and hence reduces the tolerance of the device to a radiation particle strike. This effect is called *bipolar effect* because the source-well-drain of the NMOS (PMOS) transistor act as a n-p-n (p-n-p) bipolar transistor. This effect mimics the "on" state of the parasitic bipolar transistor. With technology scaling, the channel length decreases which in turn reduces the base width (of the n-p-n transistor). Hence, this effect becomes more pronounced in scaled technologies [19, 11, 21].

Alpha-particle Source-drain Penetration (ALPEN). This charge collection mechanism results when a radiation particle strikes a MOS transistor at near-grazing incidence, such that the particle penetrates through both the source and the drain regions of the transistor. A radiation particle penetration through both the source and the drain regions of the MOS transistor (nominally in the off state) perturbs the potential in the channel region. In this case, the charge collection at the drain of the MOS transistor happens in three phases: an initial funneling phase while there is no source/drain barrier, a bipolar phase as the source/channel barrier recovers, and subsequent diffusion phase (after the device potentials have recovered). This process also mimics the "on" state of the transistor. It is reported that the charge collection due to the ALPEN mechanism increases rapidly for effective gate lengths below 0.5 μm [19, 11]. This mechanism may increase the radiation susceptibility of DSM devices.

The charge collected (through any mechanism) at the drain node of a device results in voltage transients at that node. These voltage transients in turn may result in soft errors.

1.1.1.4 Circuit Level Modeling of a Radiation Particle Strike

A radiation particle strike in a device induces current flow from the n type diffusion to the p type diffusion. Traditionally, the radiation-induced current at circuit level is modeled by a double-exponential current pulse [20] for circuit level simulations. The expression for this pulse is

$$i_{\mathrm{seu}}(t) = \frac{Q}{(\tau_\alpha - \tau_\beta)}(\mathrm{e}^{-t/\tau_\alpha} - \mathrm{e}^{-t/\tau_\beta}). \tag{1.1}$$

Here Q is the amount of charge collected as a result of the ion strike, while τ_α is the collection time constant for the junction and τ_β is the ion track establishment constant. This current pulse is injected at any node in a circuit, to simulate a radiation particle strike in SPICE at that node. Typically τ_α is of the order of 100 ps and τ_β is of the order of tens of picoseconds [12, 11]. Figure 1.2 shows $i_{\mathrm{seu}}(t)$ for several values of Q, τ_α, and τ_β. The minimum amount of charge required to result in an error is referred to as critical charge (Q_{cri}).

Note that in DSM devices, the radiation-induced current may be very different from this double exponential pulse [19, 22]. This is because, in DSM

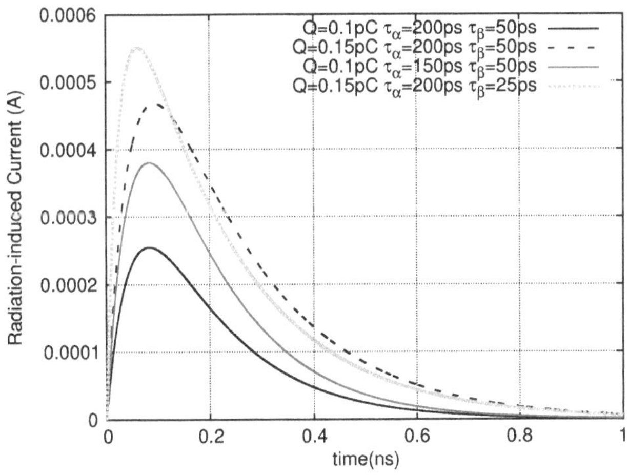

Fig. 1.2 Current pulse model for a radiation particle strike plotted for different values of Q, τ_α and τ_β

devices, the substrate is more heavily doped when compared with older technologies. As mentioned earlier, heavily-doped substrate demonstrate charge collection due to both the drift and the diffusion processes [20, 17, 18, 19, 11]. Therefore, a significant amount of charge is collected in DSM devices, due to both the drift and the diffusion processes. Whereas, in older technologies, the charge was mainly collected by the drift process. Since, the double exponential current pulse of (1.1) was derived for an older technology by using the fact that the charge is mainly collected by the drift process [20], the radiation-induced current pulse can be different from this double exponential current pulse in DSM devices. Therefore, for an accurate analysis, device-level simulations of radiation particle strikes in transistors need to be performed. However, for circuit level analysis and design, it is adequate to use the current model of (1.1) to model the worst case radiation particle strike [11, 12].

1.1.1.5 Impact of Technology Scaling on the Radiation Tolerance of VLSI Design

In the DSM era, the number of transistors on a chip is still increasing, in accordance with Moore's law [23]. This is facilitated by decreasing device and interconnect dimensions, which have led to a reduction in the node capacitances of VLSI circuits. Hence, in modern VLSI processes, even a small amount of charge deposited by a radiation particle (or low energy particle) is sufficient to cause a significant change in the voltage of a node. In other words, DSM circuits are susceptible even to low energy radiation particle strikes. This is further aggravated by decreasing supply voltages and increasing operating frequencies in the DSM regime.

Although these technology scaling trends severely reduce the radiation tolerance of VLSI circuits, there are a couple of factors associated with technology scaling which improve the radiation tolerance of VLSI circuits. The area of transistors reduces with technology scaling and hence the probability that a device in a circuit experiences a radiation particle strike reduces as well. Also, the decreasing supply voltages reduce the charge collection efficiency. Therefore, the devices implemented in newer technologies (with lower supply voltages) collect less charge compared with the devices implemented in older technologies (with higher supply voltages). A reduction in the amount of charge collected (due to lowering supply voltages) with technology scaling improves the radiation resilience of VLSI circuits. In spite of these factors, an 8% increase in soft error rate (SER) is expected per logic state bit for each technology generation [6, 7].

The soft error rate (SER) is typically measured as failure in time (FIT), where a FIT is defined as the number of failures in 10^9 hours of operation. Figure 1.3 shows the SER for an Alpha [24] processor, which was implemented using different technology nodes [4]. Figure 1.3 shows the individual contributions of SRAMs, latches (for different pipeline depths) and combinational logic (for different pipeline depths with a fanout factor of 4) to the overall SER of the Alpha processor. Observe from Fig. 1.3 that the overall chip SER, which is the sum of the contributions of SRAMs, latches, and combinational logic, increases with decreasing feature sizes.

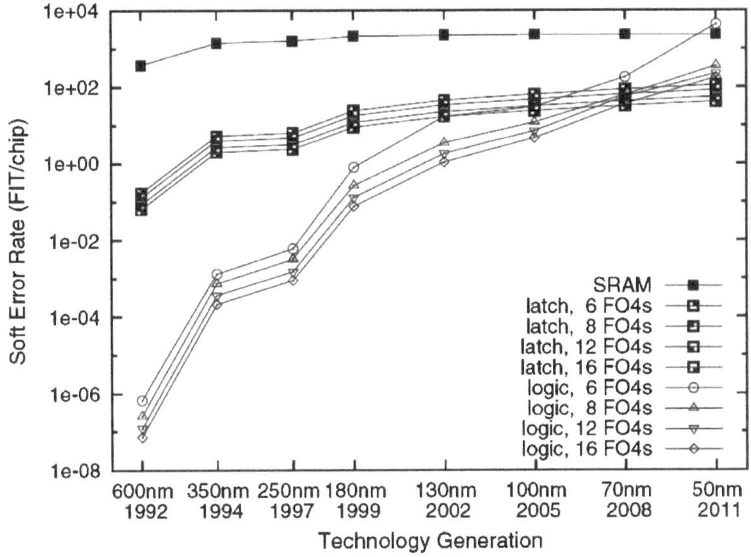

Fig. 1.3 SER of an alpha processor for different technology nodes [4]

This verifies that radiation particle strikes are becoming increasingly problematic for the reliability of VLSI systems, as predicted by the theory.

Also, observe from Fig. 1.3 that in older technologies the contribution of the SRAMs and latches to the overall chip SER was much higher than that of combinational logic. Hence, traditionally, radiation particle strikes were mainly considered problematic for memories (SRAMs, DRAMs, and latches) only. However, as the feature size is reduced below 45 nm, the SER contribution of combinational logic has increased by a large factor (more than 10^9), whereas the SER contribution of SRAMs (in absolute terms) has stayed relatively constant (as shown in Fig. 1.3). This is because of the fact that with technology scaling, heavily pipelined circuits are increasingly used, which leads to a reduction in the depth of combinational circuits. Because of this, the effect of the three masking factors (as described earlier) reduces and hence, fewer SET events are masked. Hence, it is expected that radiation particle strikes in combinational logic will be more problematic than in memories in future technologies [4, 11, 12]. Note that the SER of the Alpha processor due to radiation particle strikes in latches also increased slightly with decreasing feature sizes. Therefore, it will be necessary to harden both combinational logic and memories, to improve the radiation resilience of VLSI systems implemented using future DSM processes.

Many critical applications such as space, military, and critical terrestrial electronics (e.g., biomedical circuits and high performance servers) electronics place a stringent demand on reliable circuit operation. Therefore, efficient analysis and design techniques are required to harden VLSI circuits (both combinational logic and memories) against radiation events. Developing these is one of the two goals of this monograph.

1.1.2 Process Variations

Another important problem encountered with technology scaling in the DSM era is the increase in process variations. With the continuous scaling of devices and interconnects, variations in key device and interconnect parameters such as channel length (L), threshold voltage (V_T), oxide thickness (T_{ox}), wire width (W_M), and wire height (H) are increasing at an alarming rate [8, 9, 10]. Because of this, the performance of different die of the same IC can vary widely, resulting in a significant yield loss, which translates into higher manufacturing costs.

The two major sources of variability in device parameters are (a) limited control over the manufacturing process (extrinsic causes of variations) and (b) fundamental atomic-scale randomness of the device (intrinsic causes of variations) [8]. The variability that arises due to limited control over the manufacturing process is becoming more and more challenging to control. This is because of the inability of the semiconductor industry to improve manufacturing tolerances at the same pace as technology scaling [8]. For example, the light source (with a wavelength of 193 nm) used in lithography in older technologies ($\geq$130 nm) is still used in newer technologies (45 nm and below). Therefore, it is becoming increasingly difficult to control the channel length of transistors with technology scaling [8]. The intrinsic causes of variations are also expected to be significantly problematic in future technologies because of the fact that device dimensions are approaching the scale of silicon lattice distances. At this scale, quantum physics needs to be used to explain device operation, which is modeled as a stochastic process. Also, at this scale, the precise atomic configuration of the material significantly affects the electrical properties of the device. Therefore, a small variation in the silicon structure has a large impact on the device performance. For example, the threshold voltage of a transistor heavily depends on the doping density of the channel region. With technology scaling, the number of dopant atoms required to achieve the desired doping density is getting smaller [8]. Since the placement of dopant atoms in the silicon crystal structure is random, the final number of dopant atoms deposited in the channel region of a transistor is a random variable. Therefore, the threshold voltage of transistors also become a random variable. Variations in interconnect parameters are mainly caused by a limited control over the manufacturing process. Processing steps such as chemical mechanical polishing (CMP) and etching induce variations in interconnects or wire dimensions [8].

The process variations due to these sources can be classified as *systematic variations* and *random variations* [8, 9, 25]. The systematic component is the predictable variation trend across a chip, and is caused by spatial dependencies of device processing, such as CMP variations [26] and optical proximity effects [27]. The random component is caused by effects such as random fluctuations of the number and location of dopants in the MOSFET channel, polysilicon gate line-edge roughness, etc. [8, 9, 10].

Note that in terms of delay variability of a circuit, the contribution of variations in device parameters dominates that of interconnect parameter variations [8]. The variation in device parameters contributes close to 90% of the total variability of the delay of a realistic design [8]. In future technologies, it is expected that the variation in device parameters will continue to be the dominant source of delay variability of a circuit.

1.1.2.1 Impact of Technology Scaling on Process Variations

Figure 1.4 shows the standard deviation of the threshold voltage of transistors (σ_{V_T}) implemented in different technology nodes [28]. As shown in Fig. 1.4, σ_{V_T} has increased by a factor of ~2× for a 45 nm technology compared with a 130 nm process. Note that the absolute value of V_T is higher for 130 nm process (~0.35 V) compared with a 45 nm process (~0.28 V). Similarly, the variation in other device (L and T_{ox}) and interconnect (W, H, etc) parameters has also increased with technology scaling, as reported in [29]. Therefore, unless significant advancements are made in process control, the variation in key device and interconnect parameters is expected to further increase in future technologies.

Additionally, as devices are scaled below 45 nm, the random component of the total variations becomes significantly more problematic than the systematic component [30, 8]. Negligible spatial correlation was observed in the L and V_T of devices in a test chip fabricated using a 65 nm SOI process [30]. However, the random component of L and V_T variation was quite high in comparison (the standard deviation of L and V_T variations was 5% and 9% of the mean value, respectively). Thus, the L [30, 8, 26] and V_T [30, 8, 31, 10] variations are expected to be mostly random (or independent) in nature for future DSM technologies.

With the increasing amount (σ/μ) of variations in device and interconnect parameters, it becomes difficult to predict the performance of VLSI designs, and hence it becomes a challenging task to design reliable VLSI systems. The second goal of this monograph is to develop efficient analysis and design techniques to address the process variation issue, in order to facilitate the implementation of process variation resilient VLSI circuits. These techniques help improve design yield and hence lower manufacturing costs.

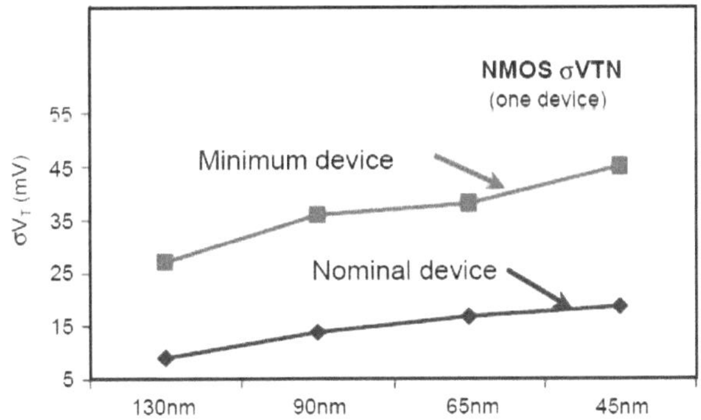

Fig. 1.4 Variation in threshold voltage of devices for different technology nodes

1.2 Monograph Overview

Section 1.1 indicates that radiation particle strikes and process variations can significantly degrade the reliability of VLSI systems. Because of the widespread use of modern VLSI circuits, it is necessary to address these issues while designing VLSI systems, so as to improve their reliability. Therefore, there is a critical need for analysis and design techniques to enable the implementation of VLSI systems that are resilient to radiation and process variation effects.

The goal of this monograph is to develop several analysis and design techniques to achieve circuit resilience against radiation particle strikes and process variations. This monograph consists of two parts.

In the first part of this monograph (Chaps. 2–7), four analysis approaches for analyzing the effects of radiation particle strikes in combinational circuits, SRAMs, and voltage scaled circuits [32, 33, 34] are presented. Two circuit level hardening approaches [35, 36] are also presented, to harden combinational circuits against a radiation particle strike.

In the second part of this monograph (Chaps. 8–10), a sensitizable statistical timing analysis approach is presented to improve the accuracy of statistical timing analysis of combinational circuits. Two design approaches are also presented to improve the process variation tolerance of combinational circuits and voltage level shifters (which are used in circuits with multiple interacting supply domains), respectively.

This monograph is organized as follows.

In Chap. 2, an analytical approach is developed to analyze the radiation-induced transients in combinational circuits. Efficient and accurate models for radiation-induced transients are required to evaluate the radiation tolerance of a circuit. As mentioned earlier, a radiation particle strike at a node may result in a voltage glitch. The pulse width of this voltage glitch is a good measure of radiation robustness of a design. Thus, an analytical model to estimate the pulse width of the radiation-induced voltage glitch in combinational designs is presented in this chapter. In this approach, a piecewise linear transistor I_{DS} model is used (instead of a linear RC gate model), and the effect of the ion track establishment constant (τ_β) of the radiation-induced current pulse is considered. Both these factors improve the accuracy (in comparison with the best existing approach [37]) of the analytical model for the pulse width computation. The model is applicable to any logic gate, with arbitrary gate size and loading, and with different amounts of charge collected due to the radiation strike. The model can be used to quickly (1,000× faster than SPICE [38]) determine the susceptible gates in a design (the gates where a radiation particle strike can result in a voltage glitch with a positive pulse width). The most susceptible gates can then be protected using circuit hardening approaches, based on the degree of hardening desired.

In Chap. 3, an analytical model is presented, which efficiently estimates the shape of the voltage glitch that results from a radiation particle strike. A model for the load current $I^G_{out}(V_{in}, V_{out})$ of the output terminal current of the gate G is used. Again, the model is applicable to any general combinational gate with different loading, and for arbitrary values of collected charge (Q). The effect of the ion track establishment constant (τ_β) of the radiation particle induced current pulse is accounted for. The voltage glitch estimated by this analytical model can be propagated to the primary outputs of a circuit using existing voltage glitch propagation tools. The properties of the voltage glitch (such as its magnitude, glitch shape and width) at the primary outputs can be used to evaluate the SEE robustness of the circuit. On the basis of the result of this analysis, circuit hardening approaches can be implemented to achieve the level of radiation tolerance required.

Chapter 4 presents a model for the dynamic stability of an SRAM cell in the presence of a radiation particle strike. Such models are required since SRAM stability analysis is crucial from an economic viewpoint, given the extensive use of memory in modern processors and SoCs. Static noise margin (SNM)-based stability SRAM stability analysis often results in pessimistic designs because SNM cannot capture the transient behavior of the noise. Therefore, to improve analysis accuracy, a dynamic stability analysis is required. The model proposed in this chapter utilizes the double exponential current pulse of (1.1) for modeling a radiation particle strike, and is able to predict (more accurately than the most accurate prior approach [39]) whether a radiation particle strike will result in a state flip in a 6T-SRAM cell (for given values of Q, τ_α and τ_β). This model enables a designer to quickly (2,000× faster than SPICE) and accurately analyze SRAM stability during the design phase.

In Chap. 5, an analysis of the effects of voltage scaling on the radiation tolerance of VLSI systems is presented. For this analysis, 3D simulations of radiation particle strikes on the output of an inverter (implemented using DVS and subthreshold design) were performed. The radiation particle strike on an inverter was simulated using Sentaurus-DEVICE [40] for different inverter sizes, inverter loads, the supply voltage values (VDD) and the energy of the radiation particles. From these 3D simulations, several nonintuitive observations were made, which are important to consider during radiation hardening of such DVS and subthreshold circuits. On the basis of these observations, several guidelines are proposed for radiation hardening of such designs. These guidelines suggest that traditional radiation hardening approaches need to be revisited for DVS and subthreshold designs. A charge collection model for DVS circuits is also proposed, using the results of these 3D simulations. The parameters of this charge collection model can be included in transistor model cards in SPICE, to improve the accuracy of SPICE-based simulations of radiation events in DVS circuits.

Chapter 6 presents a radiation tolerant combinational circuit design approach, which is based on diode clamping action. This diode clamping-based hardening approach is based on the use of shadow gates, whose task it is to protect the primary gate in case it experiences a radiation strike. The gate to be protected is duplicated locally, and a pair of diode-connected transistors (or diodes) is connected between the outputs of the original and the shadow gate. These diodes turn on when the voltage across the two gate outputs deviates (during a radiation strike). A methodology is also presented to protect specific gates of the circuit based on electrical masking, in a manner that guarantees radiation tolerance for the entire circuit and also keeps the area and delay overhead low. An improved circuit level hardening algorithm is also proposed, to further reduce the delay and area overhead. Note that the diode clamping-based approach is suitable for hardening a circuit against low energy particle strikes.

In Chap. 7, another radiation tolerant combinational circuit design approach is presented, which is called the *split-output based hardening approach*. This hardening approach exploits the fact that if a gate is implemented using only PMOS (NMOS) transistors then a radiation particle strike can result only in logic 0–1 (1–0) transient. Based on this observation, radiation hardened variants of regular static CMOS gates are derived. Split-output based radiation hardened gates exhibit an extremely high degree of radiation tolerance, which is validated at the circuit level. Hence, this approach is suitable for hardening against medium and high energy radiation particles. Using split-output gates, circuit level hardening is performed based on logical masking, to selectively harden those gates in a circuit, which contribute maximally to the soft error failure of the circuit. The gates whose outputs have a low probability of being logically masked are replaced by their radiation tolerant counterparts, such that the digital design achieves a soft error rate reduction of a desired amount (typically 90%). The split-output based hardening approach is able to harden combinational circuits with a modest layout area and delay penalty.

Chapter 8 presents the sensitizable statistical timing analysis (StatSense) methodology, developed to remove the pessimism due to two sources of inaccuracy which plague current statistical static timing analysis (SSTA) tools. Specifically, the StatSense approach implicitly eliminates false paths, and also uses different delay distributions for different input transitions for any gate. StatSense consists of two phases. In the first phase, a set of N logically sensitizable vector transitions, which result in the largest delays for a circuit, are obtained. In the second phase, these delay-critical sensitizable input vector transitions are propagated using a Monte-Carlo-based technique to obtain the delay distribution at the outputs. The specific input transitions at any gate are known after the first phase, and so the gate delay distribution corresponding to these input transitions is utilized in the second phase. The second phase performs Monte-Carlo-based statistical static timing analysis (SSTA), using the appropriate gate delay distribution corresponding to the particular input transition for each gate. The StatSense approach is able to significantly improve the accuracy of SSTA analysis. The circuit delay distribution obtained using StatSense closely matches that obtained by SPICE based Monte-Carlo simulations.

In Chap. 9, a process variation tolerant design approach for combinational circuits is presented, which exploits the fact that random variations can cause a significant mismatch in two identical devices placed next to each other on the die. In this approach, a large gate is implemented using an appropriate number (>1) of smaller gates, whose inputs and outputs are connected to each other in parallel. This parallel connection of smaller gates to form a larger gate is referred to as a *parallel gate*. Since the L and V_T variations are largely random and have independent variations in smaller gates, the variation tolerance of the parallel gate is improved. The parallel gates are implemented as single layout cells. By careful diffusion sharing in the layout of the parallel gates, it is possible to reduce the input and output capacitance of the gates, thereby improving the nominal circuit delay as well. An algorithm is also developed to selectively replace critical gates in a circuit by their parallel counterparts, to improve the variation tolerance of the circuit. Monte-Carlo simulations demonstrate that this process variation tolerant design approach achieves significant improvements in circuit level variation tolerance.

In Chap. 10, a novel process variation tolerant single-supply true voltage level shifter (SS-TVLS) design is presented. It is referred to as "true" since it can handle both low to high, or high to low voltage level conversions. The SS-TVLS is the first VLS design, which can handle both low-to-high and high-to-low voltage translation without a need for a control signal. The use of a single supply voltage reduces circuit complexity, by eliminating the need for routing both supply voltages. The proposed circuit was extensively simulated in a 90 nm technology using SPICE. Simulation results demonstrate that the level shifter is able to perform voltage level shifting with low leakage for both low to high, as well as high to low voltage level translation. The proposed SS-TVLS is also more tolerant to process and temperature variations, when compared with a combination of an inverter along with the nontrue VLS solution [41].

Finally, in Chap. 11, this monograph is concluded. This chapter also presents some future directions for research, and a summary of the broader impact of this work.

1.3 Chapter Summary

In this chapter, two major issues (radiation particle strikes and process variations), which are encountered while designing reliable VLSI systems, were introduced. With technology scaling, it is expected that the effect of these issues on the reliability of VLSI designs will become more severe. Thus, there is a critical need to address these issues while designing VLSI systems.

The next chapter will describe the first radiation analysis approach for combinational circuits.

References

1. T. May and M. Woods, "Alpha-particle-induced soft errors in dynamic memories," *IEEE Transaction on Electron Devices*, vol. ED-26, pp. 2–9, Jan 1979.
2. J. Pickle and J. Blandford, "CMOS RAM cosmic-ray-induced error rate analysis," *IEEE Transactions on Nuclear Science*, vol. NS-29, pp. 3962–3967, 1981.
3. W. Massengill, M. Alles, and S. Kerns, "SEU error rates in advanced digital CMOS," in *Proc. of the European Conf. on Radiation and Its Effects on Components and Systems*, Sep. 1993, pp. 546–553.
4. P. Shivakumar, M. Kistler, S. W. Keckler, D. Burger, and L. Alvisi, "Modeling the effect of technology trends on the soft error rate of combinational logic," in *Proc. of the Intl. Conf. on Dependable Systems and Networks*, 2002, pp. 389–398.
5. Q. Zhou and K. Mohanram, "Transistor sizing for radiation hardening," in *Proc. of the Intl. Reliability Physics Symposium*, April 2004, pp. 310–315.

6. P. Hazucha, T. Karnik, J. Maiz, S. Walstra, B. Bloechel, J. Tschanz, G. Dermer, S. Hareland, P. Armstrong, and S. Borkar, "Neutron soft error rate measurements in a 90-nm CMOS process and scaling trends in SRAM from 0.25-um to 90-nm generation," in *International Electron Devices Meeting*, Dec. 2003, pp. 21.5.1–21.5.4.
7. S. Borkar, "Designing reliable systems from unreliable components: the challenges of transistor variability and degradation," *IEEE Micro*, vol. 25, no. 6, pp. 10–16, Nov.-Dec. 2005.
8. M. Orshansky, S. R. Nassif, and D. Boning, *Design for manufacturability and statistical design: A constructive approach*, US Springer, 2008.
9. K. Agarwal and S. Nassif, "Characterizing process variation in nanometer CMOS," in *Proc. of the Design Automation Conf.*, June 2007, pp. 396–399.
10. K. Bernstein, D. J. Frank, A. E. Gattiker, W. Haensch, B. L. Ji, S. R. Nassif, E. J. Nowak, D. J. Pearson, and N. J. Rohrer, "High-performance CMOS variability in the 65-nm regime and beyond," *IBM Journal of Research and Development*, vol. 50, pp. 433–449, July/Sept. 2006.
11. P. E. Dodd and L. W. Massengill, "Basic mechanisms and modeling of single-event upset in digital microelectronics," *IEEE Transactions on Nuclear Science*, vol. 50, no. 3, pp. 583– 602, 2003.
12. Q. Zhou and K. Mohanram, "Gate sizing to radiation harden combinational logic," *IEEE Transactions on Computer-Aided Design of Integrated Circuits and Systems*, vol. 25, no. 1, pp. 155–166, Jan. 2006.
13. D. Binder, C. Smith, and A. Holman, "Satellite anomalities from galactic cosmic rays," *IEEE Transactions on Nuclear Science*, vol. NS-22, pp. 2675–2680, Dec. 1975.
14. R. G. Harrison and D. B. Stephenson, "Detection of a galactic cosmis ray influence on clouds," *Geophysical Research Abstracts*, vol. 8, no. 07661, pp. 1, 2006.
15. G. Cellere, A. Paccagnella, A. Visconti, and M. Bonanomi, "Soft errors induced by single heavy ions in floating gate memory arrays," in *Proc. of the Intl. Symposium on Defect and Fault Tolerance in VLSI Systems*, 2005, pp. 275–284.
16. A. Johnston, "Scaling and technology issues for soft error rate," in *Proc. of the Annual Research Conf. on Reliability*, Oct. 2000, pp. 1–8.
17. L. D. Edmonds, "A simple estimate of funneling-assisted charge collection," *IEEE Transactions on Nuclear Science*, vol. 38, no. 2, pp. 828–833, Apr 1991.
18. P. E. Dodd, F. W. Sexton, and P. S. Winokur, "Three-dimensional simulation of charge collection and multiple-bit upset in Si devices," *IEEE Transactions on Nuclear Science*, vol. 41, pp. 2005–2017, 1994.
19. P. E. Dodd, "Device simulation of charge collection and single-event upset," *IEEE Transactions on Nuclear Science*, vol. 43, no. 2, pp. 561–575, Apr. 1996.
20. G. Messenger, "Collection of charge on junction nodes from ion tracks," *IEEE Transactions on Nuclear Science*, vol. 29, no. 6, pp. 2024–2031, 1982.
21. O. A. Amusan, "Analysis of single event vulnerabilities in a 130 nm CMOS technology," M.S. thesis, Vanderbilt University, 2006.
22. S. DasGupta, "Trends in single event pulse widths and pulse shapes in deep submicron CMOS," M.S. thesis, Vanderbilt University, 2008.
23. G. E. Moore, "Cramming more components onto integrated circuits," *Electronics*, vol. 38, no. 8, pp. 1–4, April 1965.
24. R. E. Kessler, "The Alpha 21264 microprocessor," *IEEE Micro*, vol. 19, no. 2, pp. 24–36, 1999.
25. H. Chang and S. S. Sapatnekar, "Statistical timing analysis under spatial correlations," *IEEE Transactions on Computer-Aided Design of Integrated Circuits and Systems*, vol. 24, no. 9, pp. 1467–1482, Sept. 2005.
26. L. He, A. B. Kahng, K. H. Tam, and J. Xiong, "Simultaneous buffer insertion and wire sizing considering systematic CMP variation and random Leff variation," *Proc. of the Intl. Conf. on Computer-Aided Design*, vol. 26, no. 5, pp. 845–857, May 2007.
27. K. Cao, S. Dobre, and J. Hu, "Standard cell characterization considering lithography induced variations," in *Proc. of the Design Automation Conf.*, 2006, pp. 801–804.

28. K. J. Kuhn, "Reducing variation in advanced logic technologies: Approaches to process and design for manufacturability of nanoscale CMOS," in *International Electron Devices Meeting*, Dec. 2007, pp. 471–474.
29. S. Nassif, "Delay variability: Sources, impacts and trends," in *Proc. of the Intl. Solid State Circuits Conf.*, 2000, pp. 368–369.
30. W. Zhao, Y. Cao, F. Liu, K. Agarwal, D. Acharyya, S. Nassif, and K. Nowka, "Rigorous extraction of process variations for 65nm CMOS design," in *Proc. of the European Solid State Device Research Conf.*, Sept. 2007.
31. K. Agarwal, F. Liu, C. McDowell, S. Nassif, K. Nowka, M. Palmer, D. Acharyya, and J. Plusquellic, "A test structure for characterizing local device mismatches," in *Proc. of the Symposium on VLSI Circuits*, 2006, pp. 67–68.
32. R. Garg, C. Nagpal, and S. P. Khatri, "A fast, analytical estimator for the SEU-induced pulse width in combinational designs," in *Proc. of the Design Automation Conf.*, June 2008, pp. 918–923.
33. R. Garg and S. P. Khatri, "Efficient analytical determination of the SEU-induced pulse shape," in *Proc. of the Asia and South Pacific Design Automation Conf.*, Jan. 2009.
34. R. Garg, P. Li, and S. P. Khatri, "Modeling dynamic stability of SRAMs in the presence of single event upsets (SEUs)," in *Proc. of the Intl. Symposium on Circuits and Systems*, May 2008, pp. 1788–1791.
35. R. Garg, N. Jayakumar, S. P. Khatri, and G. Choi, "A design approach for radiation-hard digital electronics," in *Proc. of the Design Automation Conf.*, July 2006, pp. 773–778.
36. R. Garg and S. P Khatri, "A novel, highly SEU tolerant digital circuit design approach," in *Proc. of the Intl. Conf. on Computer Design*, Oct. 2008, pp. 14–20.
37. K. Mohanram, "Closed-form simulation and robustness models for SEU-tolerant design," in *Proc. of the VLSI Test Symposium*, 2005, pp. 327–333.
38. L. Nagel, "Spice: A computer program to simulate computer circuits," in *University of California, Berkeley UCB/ERL Memo M520*, May 1995.
39. B. Zhang, A. Arapostathis, S. Nassif, and M. Orshansky, "Analytical modeling of SRAM dynamic stability," in *Proc. of the Intl. Conf. on Computer-Aided Design*, Nov. 2006, pp. 315–322.
40. Synopsys Inc., Mountain View, CA, *Sentaurus user's manuals*, 2007.12 edition.
41. Q. A. Khan, S. K. Wadhwa, and K. Misri, "A single supply level shifter for multi voltage systems," in *Proc. of the Intl. Conf. on VLSI Design*, Jan. 2006, pp. 1–4.

Part I
Soft Errors

Chapter 2
Analytical Determination of Radiation-induced Pulse Width in Combinational Circuits

2.1 Introduction

With technology scaling, radiation particle strikes are becoming increasingly problematic for both combinational circuits and memory elements, as described in Chap. 1. Many critical applications such as biomedical circuits, as well as space and military electronics, demand reliable circuit functionality. Therefore, the circuits used in these application must be tolerant to radiation particle strikes.

To design radiation tolerant VLSI systems *efficiently*, it is required to first analyze the nature of radiation-induced voltage transients, and the effects of radiation particle strikes both on combinational circuits and memory elements (as SRAM cells). Then, on the basis of the findings of this analysis, circuit hardening approaches can be implemented to achieve radiation resilience while satisfying area, delay, and power constraints. This chapter and the next three chapters present radiation analysis approaches developed in this monograph for analyzing the effect of a radiation particle strike in combinational circuits and SRAMs. Then two hardening approaches are presented in Chaps. 6 and 7.

Circuit hardening approaches [1, 2, 3] often employ selective gate hardening to reduce the area and delay overhead associated with radiation hardening contributors. This is achieved by only protecting those gates in a circuit which are the significant contributors to the soft error failure rate of the circuit. Hence, the radiation susceptibility of such gates has to be examined to evaluate the radiation tolerance of the circuit. Also, for efficient hardening, it is important to harden the circuit early in the design flow. This will help in reducing the number of design iterations and reducing design turn-around time and cost. However, this can be achieved only if radiation analysis techniques can quickly and accurately simulate the effects of radiation events of different particle energies, for different gates with different loading conditions. For this, it is important to evaluate the radiation tolerance of a circuit using robustness metrics.

An exhaustive SPICE-based simulation of radiation events in a combinational circuit would be accurate; however, it would require a large number of simulations since the circuit can have a large number nodes and a radiation particle strike can occur at any one of these nodes. Also, the transient pulse resulting from a

R. Garg and S. P. Khatri, *Analysis and Design of Resilient VLSI Circuits: Mitigating Soft Errors and Process Variations*, DOI 10.1007/978-1-4419-0931-2_2,

radiation particle strike depends upon the node (node capacitance and the sizing characteristics of the gate driving that node), the amount of charge collected due to the particle strike, and the state of the circuit inputs. Therefore, it is computationally intractable to use exhaustive SPICE-based simulators for simulating the effect of radiating event in the early stages of the design flow. Thus, there is a need for efficient and accurate analytical models for SET events in combinational circuits.

The modeling of radiation events in either combinational or sequential circuits involves solving nonlinear differential equations. Because of this, not much success has been achieved in developing accurate and efficient models, which are applicable across different scenarios (such as different gate sizes, dumped charge, fanout loading, etc.). Modeling approaches in the past (explained in Sect. 2.2) have made several assumptions and approximations which limit the applicability of the resulting model due to the large error involved.

In this chapter, an analytical model for the pulse width[1] of the radiation-induced voltage glitch in combinational circuits is presented. The pulse width of the voltage glitch due to a radiation particle strike is a good measure of radiation robustness because, on the one hand, if a gate is more susceptible to radiation particle strikes, then a particle strike at the output node of that gate would result in a voltage glitch with a larger pulse width. On the other hand, if a gate is less susceptible to radiation events, then pulse width of the voltage glitch will be lower. Hence, the pulse width of the radiation-induced voltage glitch is often used as the radiation robustness metric of choice.

The model for the pulse width of the radiation-induced voltage glitch presented in this chapter uses a piecewise linear transistor I_{DS} model (instead of a linear RC gate model as done in previous approaches [4, 5]), and also considers the effect of the ion track establishment constant (τ_β) of the radiation-induced current pulse (1.1). Both these factors improve the accuracy of the analytical model for the pulse width computation. The proposed model is applicable to any logic gate, with arbitrary gate size and loading, with different amounts of charge collected due to the radiation strike. The computation of the pulse width of the voltage glitch at a gate using the model presented in this chapter is very fast and accurate; therefore, it can be easily incorporated in a design flow to implement radiation tolerant circuits. The proposed model can be used to quickly determine whether the gates in a design experience a positive pulse width as a consequence of a radiation strike. Such gates can be upsized to harden them, while accounting for logical masking [1]. This flow can be iterated until the required tolerance against radiation particle strikes is achieved.

Note that previous approaches [4, 5] neglected the contribution of the ion track establishment constant (τ_β) to simplify their model. However, in [6], it was mentioned that if the circuit response is faster than the time constants of the radiation event, then the shape of the radiation-induced current pulse is critically important to accurately model the radiation particle strike. For a 65 nm PTM [7] model card, the

---

[1] The pulse width of the radiation-induced voltage glitch is computed as the width of the voltage glitch, measured at half the supply voltage.

delay of a minimum size inverter driving a fanout of three minimum size inverters is about 13 ps, which is much smaller than the typical time constants[2] associated with a radiation particle strike. Therefore, neglecting the contribution of the τ_β term of the current pulse of (1.1) will lead to an inaccurate analysis. Through experiments it was found that ignoring τ_β results in an under-estimation of the pulse width of the radiation-induced voltage glitch by 10%. Therefore, neglecting the contribution of the τ_β term of the current pulse of (1.1) effectively diminishes the severity of the radiation particle strike, and hence leads to an optimistic estimate for the voltage glitch. Thus, it is important to consider τ_β for an accurate analysis. The model presented in this chapter considers the contribution of τ_β.

In the remainder of this chapter, Sect. 2.2 briefly discusses related previous work on modeling radiation-induced transients in combinational circuits. The model for the pulse width of the radiation-induced voltage glitch developed in this monograph is described in Sect. 2.3. Experimental results are presented in Sect. 2.4, followed by a chapter summary in Sect. 2.5.

2.2 Related Previous Work

A significant amount of work has been done on the simulation and analysis of radiation particle strikes in both combinational and sequential circuit elements [8, 9, 10, 4, 11, 12, 5]. Most of this work can be classified under one of three categories: device-level, circuit-level, and logic-level.

Device-level simulation approaches involve solving device physics equations to evaluate the effect of a radiation particle strike. In [13], three-dimensional numerical simulation is used to study the charge collection mechanism in silicon n^+/p diodes. In [14], device level three-dimensional simulation was performed to study the charge collection mechanism and voltage transients from angled ion strikes. Although device-level approaches result in very accurate analysis, they are extremely time-consuming in nature. Hence, these approaches cannot be used for large circuits.

For circuit-level and logic-level simulation approaches, a double exponential current pulse (1.1) is used to model a particle strike [15, 12, 1]. Logic-level approaches [11, 16] are utilized when the accuracy of the analysis is not very important but the speed of the analysis is very important. In these approaches, the electrical effect of radiation-induced transient is abstracted into logic-level models, which are then used in gate-level timing simulations to propagate the effects of a radiation particle strike to the memory elements at the primary outputs of the circuit. The high level of inaccuracy of these approaches makes them unattractive for robustness evaluation of circuits under radiation particle strikes.

Circuit-level simulation approaches provide accuracy and runtimes, which are intermediate between device and logic level methods. An exhaustive SPICE-based

---

[2] Typical rise times of a radiation-induced current pulse are in the range of 10–50 ps and fall times are of the order of 100 ps [6, 1].

simulation of radiation events in a circuit would be relatively accurate; however, it is still very time consuming since a large number of simulations are required to be performed due to the reasons mentioned in Sect. 2.1. Several approaches have been proposed to model radiation-induced transients in combinational circuits [17, 18, 5, 4]. In [17], the authors presented a methodology to analyze compound noise effects in circuits. Their approach utilizes look-up tables and a database generated from SPICE simulations of all the cells in a library. Many approaches [18, 5, 4] attempt to solve a nonlinear differential equation (this equation is called a Ricatti differential equation) of the transistor to obtain a closed-form reduced model for the radiation-induced transients. For this, several approximations were made in these approaches which result in a large error.

The authors of [18] presented an exact solution of the Ricatti equation using a computationally expensive infinite power series solution. In [5], a switch-level simulator is presented, where radiation-induced transient simulation is performed in two steps. In the first step, a first order RC model is used to compute the pulse width due to a radiation particle strike, and then in the second step, a set of rules are used for the propagation of the transient pulse through simple CMOS circuit blocks. Electrical-level simulations are performed to obtain pulse widths for given resistance (R_g) and capacitance (C_g) values that model a gate. Then the pulse width for other R and C values are obtained by using the linear relationships between the pulse width obtained for R_g and C_g, and the new R and C values. One major drawback of this approach is that it cannot be used for different values of the radiation-induced current parameters (Q, τ_α and τ_β). In [4], a closed-form model is reported for radiation-induced transient simulation for combinational circuits. Again, a linear RC gate model is used, which is derived using a SPICE-based calibration of logic gates for a range of values of fanout, charge collected and gate size. In [4, 5], the circuit simulation approaches *assume a linear RC gate model*, which leads to higher inaccuracy. In the DSM era, a gate cannot be accurately modeled by a linear RC model [19]. This will also be demonstrated through an experiment in Sect. 2.3. Also, these approaches neglect the contribution of the ion track establishment constant (τ_β) of the radiation-induced current pulse of (1.1), which further increases the inaccuracy of the analysis, as explained in Sect. 2.1. In contrast to these approaches, the model proposed in this chapter uses a piecewise linear transistor I_{DS} model and also considers the effect of τ_β. Both these factors improve the accuracy of the analytical model for the pulse width computation.

2.3 Proposed Analytical Model for the Pulse Width of Radiation-induced Voltage Glitch

This section describes the analytical model for the pulse width of the radiation-induced voltage glitch developed in this monograph. Section 2.3.1 discusses the effect of a radiation particle strike at the output of an inverter, using SPICE [20] simulations. The inverters used in this discussion were implemented using a 65 nm

PTM [7] model card with VDD = 1 V. The radiation-induced transient are classified into four cases, in Sect. 2.3.2. The proposed model for the pulse width computation, based on these cases, is introduced in Sect. 2.3.3. Section 2.3.4 provides the derivation of the expression for the pulse width of the radiation-induced voltage glitch.

2.3.1 Radiation Particle Strike at the Output of an Inverter

Consider an inverter INV1 driving three identical inverters as shown in Fig. 2.1a. These inverters were implemented using a 65 nm PTM [7] model card with VDD = 1 V. Note that these inverters were designed such that the switching threshold (V_{ST}) is VDD/2 (i.e., 0.5 V). Let node *a* be at logic value 0 when a radiation

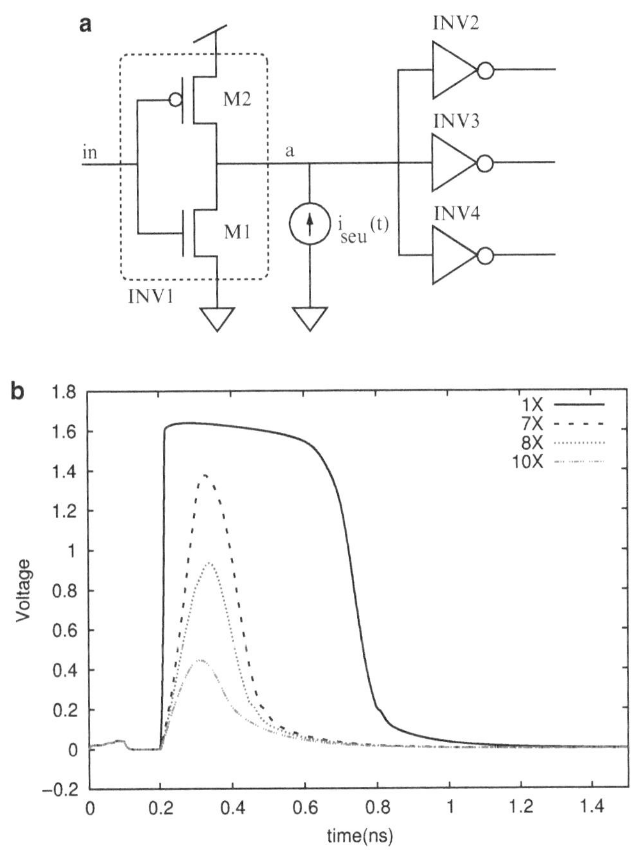

Fig. 2.1 (**a**) Radiation-induced current injected at the output of inverter INV1, (**b**) Voltage glitch at node *a*

particle strikes the diffusion of INV1. This is modeled by the injection of $i_{seu}(t)$ (described by (1.1)) at node a. The voltage glitches that results from the radiation particle strike are shown in Fig. 2.1b, for four different inverter sizes (1×, 7×, 8×, and 10×)[3] and for $Q = 150\,\text{fC}$, $\tau_\alpha = 150\,\text{ps}$ and $\tau_\beta = 50\,\text{ps}$. Note that all four inverters of Fig. 2.1a are identical.

Note that from Fig. 2.1b, in case of 10× inverters, the radiation particle strike changes the node voltage by less than V_{ST} (which is designed to be VDD/2) and hence the logic value does not change. Hence the radiation particle strike does not cause any error in circuit operation in this case. In case of the 8× inverter, the node voltage at a rises to a value around 0.9 V. As the voltage of node a starts rising, the NMOS transistor M1 of INV1 is in the linear region of operation. When the node voltage reaches V_{dsat}^N, M1 enters the saturation region of operation. For this case, the PMOS transistor is always in cut-off (since its $V_{GS} = 0$). When the radiation particle strike occurs at the output node diffusion of the 7× inverter, the magnitude of the voltage glitch is around 1.4 V. In this case as well, M1 starts out in the linear-region, and enters the saturation region when the node voltage at a rises above V_{dsat}^N. However, in this case, the PMOS transistor M2 of INV1 also turns on (in saturation mode) when the voltage of node a reaches VDD + $|V_{TP}|$ (here V_{TP} is the threshold voltage of the PMOS transistor) because the V_{GS} of M2 becomes smaller than V_{TP}. In case of the 1× inverter, the diode between the source diffusion and the bulk of M2 also turns on (M1 and M2 both conduct in the saturation region under this condition) when the voltage of node a reaches a value greater than VDD + V_{diode} (i.e., 1.6 V). Here, V_{diode} is the diode turn-on voltage which is 0.6 V for Silicon. Therefore, the voltage of node a gets clamped to a value around 1.6 V. On the basis of the above discussion, note that M1 and M2 operate in different modes of operation (cut-off, linearand saturation) during the radiation-induced transient. Therefore, *it will not be accurate to model INV1 by a linear RC gate model*, as in the case [4, 5].

Based on the above discussion, note also that the inverters of four different sizes operate quite differently during the radiation-induced transient, and the maximum voltage glitch magnitude (V_{GM}) determines their behavior at different times during the transient. In fact, when a radiation particle strikes the output node of INV1, there are four cases to consider. In the next section, each of these cases (distinguished based on the V_{GM} value) are described. Based on this classification, the proposed analytical closed form expression for the pulse width of the radiation-induced voltage glitch is derived in Sect. 2.3.4.

2.3.2 Classification of Radiation Particle Strikes

The analysis presented in this chapter is for an inverter with its input at VDD and its output at GND. The radiation particle strike results in a positive voltage glitch at the output of the gate. However, without loss of generality, the same analysis and

---

[3] The width of the NMOS (PMOS) transistor of 1× inverter is 65 nm (195 nm). The channel length of both the NMOS and PMOS transistors is 65 nm.

the same analytical model can be used for any type of gate (NAND, NOR, etc.), and for any logic values applied to its inputs. Handling of NAND, NOR, etc. gates is achieved by constructing an equivalent inverter for the gate. The size of this inverter depends on the given input values of the gate. The applicability of the proposed model to different gates was verified by applying the model to a 2-input NAND gate (for all four input combinations). These results are presented in Sect. 2.4. Note that for multiple input gates, the radiation particle strike at intermediate nodes of the gate were not considered, because the worst-case transient occurs when the particle strike occurs at the output node of the gate.

Again consider the inverter INV1 of Fig. 2.1a. INV1 can operate in 4 different *cases* during the radiation event transient, based on the maximum voltage glitch magnitude V_{GM}. The value of V_{GM} depends upon the sizes of the devices M1 and M2, the gate loading at the output node a and the value of Q, τ_α and τ_β. The pulse width of the voltage glitch is computed differently for these cases, because of the different behavior of M1 and M2 (Fig. 2.1a) for these cases. The classification of the different cases is as follows.

- *Case 1* – $V_{\mathrm{GM}} \geq \mathrm{VDD} + V_{\mathrm{diode}}$: In this case, with the increasing voltage of node a (V_{a}), M1 starts conducting in the linear region and enters the saturation region when the V_{a} becomes more than V_{dsat}^{N}. M2 starts conducting in the saturation mode once V_{a} crosses $\mathrm{VDD} + |V_{\mathrm{TP}}|$. Eventually when V_{a} reaches $\mathrm{VDD} + V_{\mathrm{diode}}$, the voltage between the source diffusion and the bulk terminal of the PMOS transistor M2 becomes $\geq V_{\mathrm{diode}}$. Therefore, the diode between these two terminals get forward biased and it starts conducting heavily. Thus V_{a} gets clamped to a value around $\mathrm{VDD} + V_{\mathrm{diode}}$.
- *Case 2* – $\mathrm{VDD} + |V_{\mathrm{TP}}| \leq V_{\mathrm{GM}} < \mathrm{VDD} + V_{\mathrm{diode}}$: In this case as well, both M1 and M2 conduct similar to Case 1. However, the diode between the diffusion and the bulk terminals of M2 remains off.
- *Case 3* – $\mathrm{VDD}/2 \leq V_{\mathrm{GM}} < \mathrm{VDD} + |V_{\mathrm{TP}}|$: Only M1 conducts in this case. M1 starts conducting in the linear region and when V_{a} crosses V_{dsat}^{N}, M1 enters the saturation region. M2 remains off in this case.
- *Case 4* – $V_{\mathrm{GM}} < \mathrm{VDD}/2$: The voltage glitch magnitude is less than VDD/2 and hence the radiation event does not result in node voltage change of magnitude greater than VDD/2.

Also, out of these 4 cases, the radiation event causes a node voltage glitch of size greater than VDD/2 for Cases 1, 2 and 3, and thus, the analysis is presented for these cases.

2.3.3 *Overview of the Model for Determining the Pulse Width of the Voltage Glitch*

Figure 2.2a schematically illustrates a voltage glitch that results from a radiation strike at the output node a of INV1. As shown in Fig. 2.2a, the node voltage rises

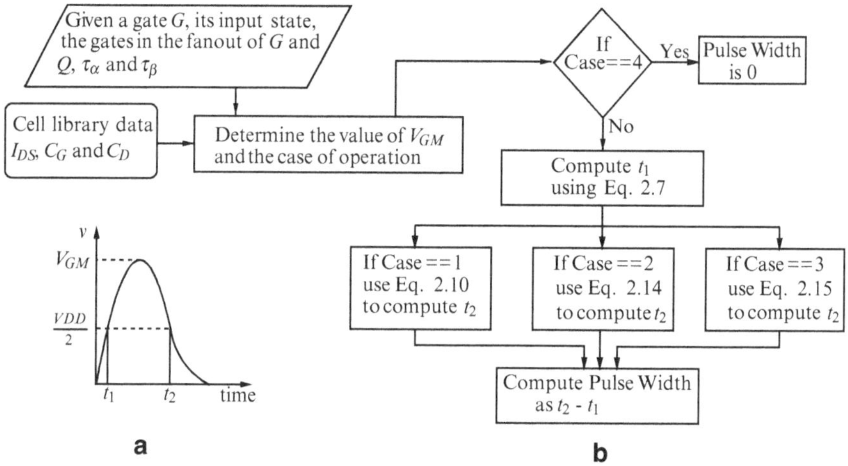

Fig. 2.2 Flowchart of the proposed model for pulse width calculation

and reaches VDD/2 at time t_1, and the node voltage falls to VDD/2 (after reaching a maximum value of V_{GM}) at the time t_2. Hence the width of the voltage glitch of Fig. 2.2a is t_2-t_1. The goal of the proposed model is to compute t_2, t_1 (the width of the glitch). To use the proposed model to compute pulse width, all the gates of different types and sizes in the library (*LIB*) need to be characterized (this was done using SPICE [20]). For each gate (for all input combinations), the current through the pull-down and pull-up stacks as a function of the gate output voltage was computed, and stored in a look-up table. The input gate capacitance (C_{G}) and the output node diffusion capacitance (C_{D}) were also computed as a function of the input (output) node voltage and stored in look-up tables. For these lookup table entries, the characterization was performed in the discrete steps of 0.1 V. For example, for INV1 of Fig. 2.1a, the drain to source current I_{DS} through M1 was computed for different V_{DS} values across M1, when node *in* is at VDD. The I_{DS} value for M2 was also computed when *in* is at the GND value, for different values of V_{DS} across M2. Thus, the number of current look-up tables (the pull-up and the pull-down current tables) for any gate is equal to 2^n (where n is the number of inputs of a gate). Similarly, C_{D} was also computed depending upon the input state of the gate. Therefore, for an n-input gate, the total size of the look-up tables for C_{G}, C_{D} and the current through the pull-down (pull-up) stack are $23 \cdot n$, $17 \cdot 2^n$, and $17 \cdot 2^n$, respectively. The saturation voltage V_{dsat} was also obtained for both NMOS and PMOS transistors, for the nominal supply voltage value. Note that the proposed model can be used for a circuit employing voltage scaling by obtaining the V_{dsat} values for different supply voltage values. The gate characterization step needs to be performed once for each gate in a library, and thus it does not affect the run-time of the model.

Figure 2.2b shows the flowchart of the algorithm used by the proposed model to compute the values of t_1 and t_2 (and hence estimate the pulse width of the voltage glitch). The input to the model is a gate G (the radiation event is to be simulated at the output node of gate G), its input state, the list of gates which are driven by the gate G, and the values of Q, τ_α and τ_β. The model first computes V_{GM} and then determines the case that is applicable. If $V_{GM} < VDD/2$ (i.e., Case 4 applies), then the pulse width is 0 else t_1 is computed. Note that the expression of t_1 is the same for cases 1, 2, and 3. After this, the time t_2 is computed using case specific expressions. Finally the pulse width of the voltage glitch $(t_2 - t_1)$ is returned. The steps of the proposed model to compute the pulse width of the voltage glitch are explained in detail in the following subsections.

2.3.4 *Derivation of the Proposed Model for Determining the Pulse Width of the Voltage Glitch*

As mentioned earlier, the discussion of the proposed model assumes that INV1 (Fig. 2.1a) has its input node *in* at VDD and the output node a at GND. A radiation particle strike results in a positive voltage glitch at node a. To ensure that the model for radiation events in combinational circuit elements is manageable, a piece-wise linear drain-source current (I_{DS}) expression was used. Consider an NMOS transistor with the input gate terminal at VDD. Then I_{DS} as a function of V_{DS} can be written as:

$$I_{DS}^{V_{DS}} = \begin{cases} V_{DS}/R_n & \text{linear } \left(V_{DS} < V_{dsat}^{N}\right) \\ K_3 + K_4 \cdot V_{DS} & \text{saturation } \left(V_{DS} \geq V_{dsat}^{N}\right). \end{cases}$$

Here, R_n is the linear region resistance, which is calculated using the I_{DS} vs. V_{DS} lookup table for V_{DS} values less than V_{dsat}^{N}. Similarly, the constants K_3 and K_4 are obtained by using the I_{DS} vs. V_{DS} lookup table, for V_{DS} values greater than V_{dsat}^{N}.

To determine the case that is applicable, it is first required to calculate the value of V_{GM}. This is done as follows.

2.3.4.1 Voltage Glitch Magnitude V_{GM}

A radiation event can result in a voltage glitch with positive pulse width only if $I_{seu}^{max} > I_{DS}^{VDD/2}$, where I_{seu}^{max} is the maximum value of radiation-induced current pulse of (1.1). This condition is used to check whether a radiation event will result in a voltage glitch of positive pulse width or not. The differential equation for the radiation-induced voltage transient at the output of INV1 of Fig. 2.1a is given by:

$$C\frac{dV_a(t)}{dt} + I_{DS}^{V_a} = i_{seu}(t), \tag{2.1}$$

where C is the capacitance[4] at node a. Equation (2.1) is accurate for values of V_a between 0 V and VDD$+|V_{\text{TP}}|$. It is used to calculate V_{GM}. Note that if the estimated V_{GM} from (2.1) is greater than VDD $+ 0.6V$, then it is assumed that Case 1 applies. In some instances, a Case 2 V_{GM} value can be diagnosed as a Case 1 situation, which results in a pessimistic pulse width estimate. The above equation can be integrated with the initial condition $V_a(t) = 0$ at $t = 0$ to obtain $V_a(t)$. For deep submicron processes, V_{dsat} is much lower than $V_{\text{GS}} - V_{\text{T}}$ due to short channel effects. For the 65 nm PTM [7] model card used in this work, V_{dsat} for both NMOS and PMOS transistors is lower than VDD/2. Therefore, to obtain the V_{GM} value, (2.1) is first integrated from the initial condition using the linear region equation for $I_{\text{DS}}^{V_a}$ till V_a reaches V_{dsat}^N value. Then, (2.1) is again integrated using the saturation region equation for $I_{\text{DS}}^{V_a}$ to obtain the $V_a(t)$ expression. The resulting expression for $V_a(t)$ is used to calculate the value of V_{GM}.

Integrating (2.1) using the linear region equation for $I_{\text{DS}}^{V_a}$ and with the initial condition $V_a(t) = 0$ at $t = 0$ gives:

$$V_a(t) = \frac{I_n}{C}\left(\frac{\mathrm{e}^{-t/\tau_\alpha}}{X} - \frac{\mathrm{e}^{-t/\tau_\beta}}{Y} - Z\mathrm{e}^{-t/R_nC}\right), \tag{2.2}$$

where

$$X = \frac{1}{R_nC} - \frac{1}{\tau_\alpha},\ Y = \frac{1}{R_nC} - \frac{1}{\tau_\beta},\ I_n = \frac{Q}{\tau_\alpha - \tau_\beta},\ Z = \frac{1}{X} - \frac{1}{Y}.$$

To obtain the time T_{sat} when $V_a(t)$ reaches the V_{dsat}^N value from (2.2), linearly expand (2.2) around the initial guess T_{sat}^a. The resulting expression for T_{sat} is:

$$T_{\text{sat}} = T_{\text{sat}}^a + \frac{V_{\text{dsat}}^N - \frac{I_n}{C}\left(\frac{\mathrm{e}^{-T_{\text{sat}}^a/\tau_\alpha}}{X} - \frac{\mathrm{e}^{-T_{\text{sat}}^a/\tau_\beta}}{Y} - Z\mathrm{e}^{-T_{\text{sat}}^a/R_nC}\right)}{\frac{I_n}{C}\left(-\frac{\mathrm{e}^{-T_{\text{sat}}^a/\tau_\alpha}}{\tau_\alpha X} + \frac{\mathrm{e}^{-T_{\text{sat}}^a/\tau_\beta}}{\tau_\beta Y} + \frac{Z}{R_nC}\mathrm{e}^{-T_{\text{sat}}^a/R_nC}\right)}. \tag{2.3}$$

To obtain the initial guess T_{sat}^a, approximate the rising part of the radiation-induced current by a line between the origin and the point where $i_{\text{seu}}(t)$ of (1.1) reaches its maximum value $I_{\text{seu}}^{\max}$. The radiation-induced current $i_{\text{seu}}(t)$ reaches $I_{\text{seu}}^{\max}$ at $T_{\text{seu}}^{\max}$. Then substitute this approximated radiation-induced current in the RHS of (2.1) and integrate it from the initial condition $V_a(t) = 0$ at $t = 0$ to $V_a(t) = V_{\text{dsat}}^N$ at $t = T_{\text{sat}}^a$ using the linear region equation for $I_{\text{DS}}^{V_a}$. After this, solve for T_{sat}^a by

[4] The value of C is obtained by the addition of the capacitance of the output diffusion node of INV1 (C_{D}), interconnect capacitance and the input capacitance of the gates driven by INV1 ($n \cdot C_{\text{G}}$). Here, n is the fanout factor. Note that these capacitance values were obtained over the operating voltage range.

performing a quadratic expansion of the resulting equation around the origin. The expression for T_{sat}^{a} is:

$$T_{\text{sat}}^{\text{a}} = \sqrt{\frac{2V_{\text{dsat}}^{N} \times C \times T_{\text{seu}}^{\max}}{I_{\text{seu}}^{\max}}}, \tag{2.4}$$

where

$$T_{\text{seu}}^{\max} = \frac{\tau_\alpha \tau_\beta}{\tau_\alpha - \tau_\beta} \log \frac{\tau_\alpha}{\tau_\beta} \quad \text{and} \quad I_{\text{seu}}^{\max} = i_{\text{seu}}(T_{\text{seu}}^{\max}).$$

So far the expression for T_{sat} is known, which is the time when $V_a(t)$ reaches V_{dsat}^{N}, or the time when M1 enters the saturation mode. Now, again integrate (2.1) with the initial condition $V_{\text{a}}(t) = V_{\text{dsat}}^{N}$ at $t = T_{\text{sat}}$, and using the saturation region current equation for $I_{\text{DS}}^{V_a}$. The resulting expression for $V_a(t)$ is:

$$V_a(t) = \frac{I_n}{C}\left(\frac{\mathrm{e}^{-t/\tau_\alpha}}{X'} - \frac{\mathrm{e}^{-t/\tau_\beta}}{Y'}\right) - \frac{K_3}{K_4} + Z'\mathrm{e}^{-K_4 t/C} \tag{2.5}$$

where,

$$X' = \frac{K_4}{C} - \frac{1}{\tau_\alpha}, Y' = \frac{K_4}{C} - \frac{1}{\tau_\beta} \quad \text{and}$$

$$Z' = V_{\text{dsat}}^{N}\mathrm{e}^{K_4 T_{\text{sat}}/C} - \frac{I_n}{C}\mathrm{e}^{K_4 T_{\text{sat}}/C}\left(\frac{\mathrm{e}^{-T_{\text{sat}}/\tau_\alpha}}{X'} - \frac{\mathrm{e}^{-T_{\text{sat}}/\tau_\beta}}{Y'}\right) + \frac{K_3}{K_4}\mathrm{e}^{K_4 T_{\text{sat}}/C}.$$

To calculate the value of V_{GM}, first differentiate (2.5) and then equate $\mathrm{d}V_a(t)/\mathrm{d}t$ to zero and solve for $T_{V_{\text{GM}}}$ (the time at which $V_a(t)$ reaches its maximum value). Since the equation $\mathrm{d}V_a(t)/\mathrm{d}t = 0$ is also a transcendental equation, hence linearly expand $\mathrm{d}V_a(t)/\mathrm{d}t = 0$ around $T_{\text{seu}}^{\max}$ and then solve for $T_{V_{\text{GM}}}$. The expression for $T_{V_{\text{GM}}}$ is:

$$T_{V_{\text{GM}}} = T_{\text{seu}}^{\max} + \frac{\frac{\mathrm{e}^{-T_{\text{seu}}^{\max}/\tau_\alpha}}{\tau_\alpha X'} - \frac{\mathrm{e}^{-T_{\text{seu}}^{\max}/\tau_\beta}}{\tau_\beta Y'} + \frac{K_4 Z'}{C}\mathrm{e}^{-K_4 T_{\text{seu}}^{\max}/C}}{\frac{\mathrm{e}^{-T_{\text{seu}}^{\max}/\tau_\alpha}}{\tau_\alpha^2 X'} - \frac{\mathrm{e}^{-T_{\text{seu}}^{\max}/\tau_\beta}}{\tau_\beta^2 Y'} + \frac{K_4^2 Z'}{C^2}\mathrm{e}^{-K_4 T_{\text{seu}}^{\max}/C}}. \tag{2.6}$$

Now, calculate V_{GM} by substituting $T_{V_{\text{GM}}}$ obtained from (2.6) in to (2.5). Note that by using this method, V_{GM} can be evaluated to be greater than VDD $+ 0.6V$, because the diode is not modeled in (2.1). Therefore, if $V_{\text{GM}} >$ VDD $+ 0.6V$ then set $V_{\text{GM}} =$ VDD $+ 0.6V$. Also note that the effect of the turning on of M2 is also not included (when $V_{\text{a}}(t)$ reaches a value above VDD $+ |V_{\text{TP}}|$). This is done to keep the analysis simple. It was found that neglecting the contribution of M2's current minimally affect the accuracy of the proposed model. The value of V_{GM} determines the case which is applicable. If Case 4 applies, then the pulse width is 0 since the radiation event does not affect the logic level of INV1. Otherwise, the times t_1 and t_2 are computed to calculate the pulse width of the voltage glitch at node a.

2.3.4.2 Derivation of the Expression for t_1

As shown in the flowchart of the proposed model in Fig. 2.2, the method to compute t_1 is identical for cases 1, 2, or 3. To obtain the expression for t_1, substitute $t = t_1$ and $V_a(t_1) = \mathrm{VDD}/2$ in (2.5) and then solve for t_1 after expanding (2.5) linearly around the point t_1^a (which is an initial guess for t_1). Here $t_1^a = T_{sat}VDD/(2V_{dsat}^N)$, which is an estimate of t_1, obtained by extrapolating along the line between (0,0) and (T_{sat}, V_{dsat}^N) The expression for t_1 is therefore:

$$t_1 = t_1^a + \frac{\frac{e^{-t_1^a/\tau_\alpha}}{X'} - \frac{e^{-t_1^a/\tau_\beta}}{Y'} + \frac{C}{I_n}\left(Z'e^{-K_4t_1^a/C} - \frac{K_3}{K_4} - \frac{\mathrm{VDD}}{2}\right)}{\frac{e^{-t_1^a/\tau_\alpha}}{X'\tau_\alpha} - \frac{e^{-t_1^a/\tau_\beta}}{Y'\tau_\beta} + \frac{K_4Z'}{I_n}e^{-K_4t_1^a/C}}. \tag{2.7}$$

Equation (2.7) gives the time at which the voltage at node a reaches $\mathrm{VDD}/2$. Note that the contribution of τ_β is not ignored in the calculation of t_1 (unlike [4, 5]).

2.3.4.3 Derivation of the Expression for t_2

The method for obtaining the value of t_2 depends upon the value of V_{GM} (i.e., the case that is applicable). The derivation of the expression for t_2, for the different cases is as follows:

Case 1. Consider the voltage and current waveforms of the 1× inverter during the radiation event as shown in Fig. 2.3. Figure 2.3 shows the voltage of node a, I_{DS} currents of M1 and M2, and the radiation-induced current pulse (i_{seu}). As shown in Fig. 2.3, when $i_{seu}(t)$ becomes equal to the I_{DS} of M1, then at that instant, the I_{DS} of M2 is approximately equal to 0 and the voltage at node a is $\mathrm{VDD} + |V_{TP}|$. This is an important observation because this information will be used as the initial condition

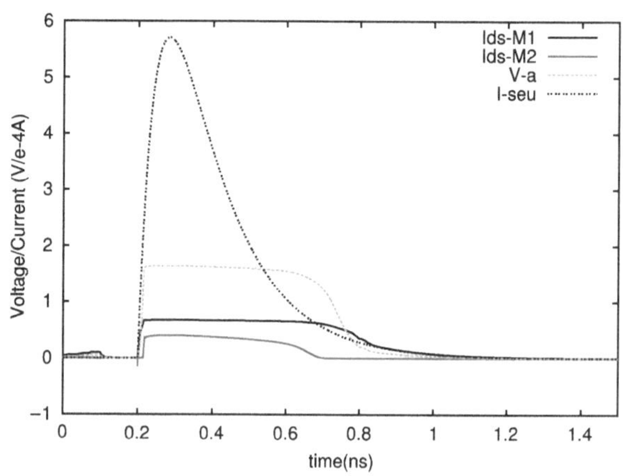

Fig. 2.3 Voltage/current due to a radiation particle strike at node a of INV1 of Fig. 2.1a

when integrating the INV1 output node voltage differential equation (2.1). Let $i_{seu}(t)$ becomes equal to the I_{DS} of M1 at time t_3. Then $V_a(t_3) = \text{VDD}+|V_{TP}|$. To calculate t_3, ignore the contribution of the e^{-t/τ_β} term of $i_{seu}(t)$. This is reasonable since τ_α is usually 3–4 times of τ_β and therefore e^{-t/τ_β} approaches 0 much faster than the e^{-t/τ_α} term. Thus the value of e^{-t/τ_β} around t_3 (which is greater than T_{seu}^{max}) will be approximately equal to 0. The expression of t_3 thus obtained by equating $i_{seu}(t)$ (ignoring the e^{-t/τ_β} term) and $I_{DS}^{\text{VDD}+|V_{TP}|}$ is:

$$t_3 = -\tau_\alpha \log \frac{I_{DS}^{\text{VDD}+|V_{TP}|}}{I_n}. \tag{2.8}$$

Now, the radiation-induced current after time t_3 is modeled by a line, one of whose end-points has a current value of $I_{DS}^{avg} = 0.5\times(I_{DS}^{\text{VDD}+|V_{TP}|} + I_{DS}^{\text{VDD}/2})$ at a time value of t_3. The other end-point has its current value as 0 at time t^*. The value of t^* is obtained by equating the charge deposited by the actual radiation-induced current $i_{seu}(t)$ from time t_3 to infinity and the charge deposited by linearized radiation-induced current equation. Hence the expression for the radiation-induced linear current model is:

$$i_{seu}^m(t) = I_{DS}^{avg}(1 - \frac{t - t_3}{t^* - t_3}) = K_1 - K_2 t, \tag{2.9}$$

where

$$t^* = t_3 + 2\frac{I_n(\tau_\alpha e^{-t_3/\tau_\alpha} - \tau_\beta e^{-t_3/\tau_\beta})}{I_{DS}^{avg}}.$$

Now substitute $i_{seu}^m(t)$ for $i_{seu}(t)$ in (2.1), use the saturation region equation for $I_{DS}^{V_a}$ and then integrate the resulting differential equation from time t_3 to t_2 (where $V_a(t_3) = \text{VDD} + |V_{TP}|$ and $V_a(t_2) = \text{VDD}/2$). The resulting equation is solved for t_2 by performing a quadratic expansion around the t_2^{a1} point. The resulting expression for t_2 is:

$$t_2 = t_2^{a1} + \frac{-Q + \sqrt{Q^2 - 4PR}}{2P}, \tag{2.10}$$

where

$$P = \frac{MK_4^2 e^{-K_4 t_2^{a1}/C}}{2C^2}, Q = \frac{K_2}{K_4} - \frac{MK_4 e^{-K_4 t_2^{a1}/C}}{C},$$

$$R = N + \frac{K_2 t_2^{a1}}{K_4} + M e^{-K_4 t_2^{a1}/C}, N = \frac{\text{VDD}}{2} - \frac{K_1 - K_3}{K_4} - \frac{K_2 C}{K_4^2},$$

$$M = e^{-K_4 t_3/C}\left(-VDD - |V_{TP}| + \frac{K_1 - K_3}{K_4} - \frac{t_3 K_2}{K_4} + \frac{K_2 C}{K_4^2}\right).$$

To obtain the value of t_2^{a1}, again integrate (2.1) but this time substitute $I_{\mathrm{DS}}^{V_a}$ by a constant current of value $I_{\mathrm{DS}}^{\mathrm{VDD}+|V_{\mathrm{TP}}|}$. The radiation-induced current is again modeled by a line with one end-point having a current value of $I_{\mathrm{DS}}^{\mathrm{VDD}+|V_{\mathrm{TP}}|}$ at a time value of t_3. The other end-point is again found by equating the charge deposited by the actual radiation-induced current $i_{\mathrm{seu}}(t)$ from time t_3 to infinity and the charge deposited by linearized radiation-induced current equation. Equation (2.1) is integrated from time t_3 to t_2^{a1}. A closed form expression can be obtained for t_2^{a1}. The resulting expression for t_2^{a1} is:

$$t_2^{a1} = t_3 + \sqrt{\frac{C\,(\mathrm{VDD}/2 + |V_{\mathrm{TP}}|)\,(t^* - t_3)}{I_{\mathrm{DS}}^{\mathrm{VDD}+|V_{\mathrm{TP}}|}}}. \tag{2.11}$$

Case 2. In this case, both M1 and M2 conduct because the magnitude of the voltage glitch is between VDD + $|V_{\mathrm{TP}}|$ and VDD + $0.6V$. Similar to Case 1, at time t_3, $i_{\mathrm{seu}}(t)$ becomes equal to $I_{\mathrm{DS}}^{\mathrm{VDD}+|V_{\mathrm{TP}}|}$ and the voltage of node a is VDD + $|V_{\mathrm{TP}}|$. The value of t_3 is again obtained using (2.8). To obtain the expression for t_2, integrate (2.1) with the initial condition $V_a(t_3) = \mathrm{VDD} + |V_{\mathrm{TP}}|$, using the saturation region current equation for the I_{DS} of M1. The resulting equation of $V_a(t)$ is:

$$V_a(t) = \frac{I_{\mathrm{n}}}{C}\left(\frac{\mathrm{e}^{-t/\tau_\alpha}}{X'} - \frac{\mathrm{e}^{-t/\tau_\beta}}{Y'}\right) - \frac{K_3}{K_4} + Z''\mathrm{e}^{-K_4 t/C}, \tag{2.12}$$

where

$$Z'' = (\mathrm{VDD} + |V_{\mathrm{TP}}|)\mathrm{e}^{K_4 t_3/C} - \frac{I_{\mathrm{n}}}{C}\mathrm{e}^{K_4 t_3/C}\left(\frac{\mathrm{e}^{-t_3/\tau_\alpha}}{X'} - \frac{\mathrm{e}^{-t_3/\tau_\beta}}{Y'}\right) + \frac{K_3}{K_4}\mathrm{e}^{K_4 t_3/C}.$$

Now use (2.12) to compute t_2. For this substitute $t = t_2$ and $V_a(t_2) = \mathrm{VDD}/2$ in (2.12), expand it around the initial guess point t_2^{a2} and then solve for t_2. Through some simulations and analysis, it was observed that t_2^{a2} (the time when $i_{\mathrm{seu}}(t)$ falls to $I_{\mathrm{DS}}^{\mathrm{VDD}/2}$ after reaching $I_{\mathrm{seu}}^{\max}$) can be used as an initial guess for t_2 since the node voltage at that time is close to VDD/2. For finding an expression for t_2^{a2}, ignore the contribution of the e^{-t/τ_β} term of $i_{\mathrm{seu}}(t)$. The expression for t_2^{a2} is:

$$t_2^{a2} = -\tau_\alpha \log \frac{I_{\mathrm{DS}}^{\mathrm{VDD}/2}}{I_{\mathrm{n}}}. \tag{2.13}$$

Now equate (2.12) to VDD/2, expand it around t_2^{a2} (from (2.13)) and then solve it for t_2. The resulting expression for t_2 is:

$$t_2 = t_2^{a2} + \frac{\frac{\mathrm{e}^{-t_2^{a2}/\tau_\alpha}}{X'} - \frac{\mathrm{e}^{-t_2^{a2}/\tau_\beta}}{Y'} + \frac{C}{I_{\mathrm{n}}}\left(Z''\mathrm{e}^{-K_4 t_2^{a2}/C} - \frac{K_3}{K_4} - \frac{\mathrm{VDD}}{2}\right)}{\frac{\mathrm{e}^{-t_2^{a2}/\tau_\alpha}}{X'\tau_\alpha} - \frac{e^{-t_2^{a2}/\tau_\beta}}{Y'\tau_\beta} + \frac{K_4 Z''}{I_n}e^{-K_4 t_2^{a2}/C}}. \tag{2.14}$$

Case 3. In this case, only M1 of Fig. 2.1a conducts because the magnitude of the glitch voltage is less than VDD + $|V_{TP}|$. Therefore, the voltage of node a from (2.5) can be used to compute t_2. The initial guess for t_2 is obtained in the same manner as Case 2 using (2.13). Now equate (2.5) to VDD/2, expand it around t_2^{a2} (from (2.13)) and then solve it for t_2. Hence the expression for t_2 is:

$$t_2 = t_2^{a2} + \frac{\frac{e^{-t_2^{a2}/\tau_\alpha}}{X'} - \frac{e^{-t_2^{a2}/\tau_\beta}}{Y'} + \frac{C}{I_n}\left(Z' e^{-K_4 t_2^{a2}/C} - \frac{K_3}{K_4} - \frac{\text{VDD}}{2}\right)}{\frac{e^{-t_2^{a2}/\tau_\alpha}}{X'\tau_\alpha} - \frac{e^{-t_2^{a2}/\tau_\beta}}{Y'\tau_\beta} + \frac{K_4 Z'}{I_n} e^{-K_4 t_2^{a2}/C}}. \qquad (2.15)$$

Using the values of t_1 and t_2 obtained in this section (for Cases 1–3), the pulse width of the radiation-induced voltage glitch at node a can be calculated. Note that τ_β is not ignored in the calculation of t_2 as well as t_1. The contribution of the e^{-t/τ_β} term of $i_{seu}(t)$ was ignored only during the *calculation of the initial guess* for t_2.

2.4 Experimental Results

The accuracy of the model proposed in this chapter for determining the pulse width of the radiation-induced voltage glitch was compared with SPICE [20]. The model was implemented in *perl* and it is much faster than SPICE simulation. In particular, for the results shown in this section, the SPICE simulations for the inverter with input 1 (input 0) took 12.6 s (10.9 s) while the perl script generated the result for input 1 as well as input 0 in 0.008 s. Thus, the proposed model is more than 1,000× faster. Note that all experiments were conducted on a Linux-based 3.6 GHz Pentium 4 machine, with 3 GB of RAM.

A standard cell library *LIB* was implemented using a 65 nm PTM [7] model card with VDD $= 1V$. The library contains INV, NAND, and NOR gates of different sizes and different numbers of inputs. As mentioned in Sect. 2.3.3, all gates in *LIB* were precharacterized. Specifically, look-up tables for the current through both the pull-up and pull-down stacks, the input gate capacitance C_G and the output node diffusion capacitance C_D (for all input combinations) were obtained for all the gates in *LIB*. The method used to obtain the stack current as well as, C_G and C_D look-up tables is explained in Sect. 2.3.3. For all experimental results reported in this section, $Q = 150$ fC, $\tau_\alpha = 150$ ps, and $\tau_\beta = 50$ ps. Similar results were obtained for the other values of Q, τ_α, and τ_β.

The proposed model was applied to inverters of different sizes (with both possible input values) for determining the pulse width of the voltage glitch induced by a radiation particle strike. The circuit under consideration is similar to Fig. 2.1a, where INV1 is driving either 1 or 3 inverters of the same size, and a radiation particle strike occurs at the output node of INV1. The results thus obtained from SPICE and the model are reported in Table 2.1. In Table 2.1, Column 1 reports the number of inverters (of the same size as INV1) present in the fanout of INV1. Column 2 reports the size of INV1 in terms of multiples of a minimum-sized inverter. Columns 3–9 report the results when the input of INV1 is at the logic value 1. Columns 3 and 4 report the values of times t_1 and t_2 obtained using SPICE. Column 5 reports the pulse width (PW^S) of the voltage glitch that results from the radiation particle strike obtained from the SPICE. Columns 6–8 report the values of t_1, t_2 and the pulse width (PW^M) calculated by the proposed model. The percentage error of the proposed model in the estimation of the pulse width, compared with SPICE, is reported in Column 9. Columns 10–16 report the same results as Columns 3–9 but for the input value of 0. As reported in Table 2.1, the proposed model estimates the pulse width of the voltage glitch due to radiation events quite accurately. The absolute average estimation error of the model is just 2.07% and 2.15% for the INV1 input values 0 and 1.

To demonstrate the applicability of the model to multiple input gates, the model was also applied to a 2-input NAND gates of different sizes (for all input combinations). The 2-input NAND gate drive either 1 or 3 inverters of the same size as the equivalent inverter of the NAND2 gate, and a radiation particle strike was assumed to occur at the output node of the NAND2 gate. The results obtained from SPICE and the model are reported in Table 2.2 for all possible input states. Note that a '–' entry in Table 2.2 means that a Case 4 situation was found (no glitch). From Table 2.2, observe that the absolute average estimation error of the model is no larger than 3.87%. For other input states, the inaccuracy of the model is even lower. The slight inaccuracy of the proposed model is due to three reasons: (1) sometimes the model wrongly diagnoses a Case 2 situation as a Case 1 as situation, as mentioned in Sect. 2.3.4.1; (2) the contribution of the capacitance of the internal node to the output node diffusion capacitance C_D in NAND2 was not accurately estimated; and (3) the Miller feedback from the output node of the loading gates (like INV2 of Fig. 2.1) to the node where a radiation particle strike affects the pulse width of the voltage glitch. In the proposed model, the effects due to this feedback were not considered. To accurately estimate the contribution of the internal node capacitance to the output node diffusion capacitance C_D in NAND2, the approach of [19] can be used to characterize NAND2 gates.

It can be concluded from Tables 2.1 and 2.2 that the proposed model for the pulse width of the voltage glitch due to a radiation event is very accurate. The worst case average estimation error for inverters and 2-input NAND gate is less than 4%. Compared with previous approaches [5, 4], the error of the proposed model is much lower. Note that these previous approaches neglected the contribution of τ_β of the radiation-induced current which leads to under-estimation of the pulse width of the voltage glitch by 10%. Hence, the inaccuracy of these previous approaches is high.

Table 2.1 Pulse width for INV1 Gate for $Q = 150\,\text{fC}$, $\tau_\alpha = 150\,\text{ps}$, and $\tau_\beta = 50\,\text{ps}$

		INV1 with input 1							INV1 with input 0						
		SPICE			Model				SPICE			Model			
Load	Size	t_1(ps)	t_2(ps)	PW^S(ps)	t_1(ps)	t_2(ps)	PW^M(ps)	% error	t_1(ps)	t_2(ps)	PW^S(ps)	t_1(ps)	t_2(ps)	PW^M(ps)	% error
1	1	7	540	533	7	540	533	0.00	7	524	517	6	529	522	0.97
1	2	12	426	414	12	427	415	0.24	11	415	404	11	421	410	1.49
1	4	22	314	292	22	319	296	1.37	20	305	285	19	317	298	4.56
1	6	33	246	213	35	258	223	4.69	30	238	208	29	261	231	11.06
1	8	50	192	142	49	195	146	2.82	44	184	140	43	184	141	0.71
3	1	10	562	552	9	563	553	0.18	9	544	535	9	542	533	−0.37
3	2	16	448	432	15	450	434	0.46	15	435	420	14	434	420	0.00
3	4	28	336	308	27	342	315	2.27	25	326	301	24	330	306	1.66
3	6	42	269	227	42	281	239	5.29	37	258	221	36	257	221	0.00
3	8	62	209	147	61	214	152	3.40	53	200	147	51	199	148	0.68
AVG								2.07							2.15

Table 2.2 Pulse width for NAND2 gate for $Q = 150\,\text{fC}$, $\tau_\alpha = 150\,\text{ps}$, and $\tau_\beta = 50\,\text{ps}$

		Inputs 11			Inputs 00			Inputs 01			Inputs 10		
Load	Size	PW^S(ps)	PW^M(ps)	% error	PW^S(ps)	PW^M(ps)	% error	PW^S(ps)	PW^M(ps)	% error	PW^S(ps)	PW^M(ps)	% error
1	1	497	501	0.80	404	402	−0.5	521	523	0.38	531	531	0.00
1	2	382	388	1.57	284	288	1.41	408	410	0.49	417	418	0.24
1	4	259	270	4.25	140	141	0.71	289	297	2.77	298	304	2.01
1	6	172	192	11.63	–	–	–	211	228	8.06	220	220	0.00
3	1	512	518	1.17	413	415	0.48	539	538	−0.19	548	548	0.00
3	2	396	404	2.02	292	300	2.74	423	424	0.24	432	434	0.46
3	4	271	285	5.17	145	145	0.0	304	310	1.97	312	318	1.92
3	6	183	191	4.37	–	–	–	224	225	0.45	232	233	0.43
AVG				3.87			0.97			1.82			0.63

2.5 Chapter Summary

With the increasing demand for reliable systems, it is necessary to design radiation tolerant circuits efficiently. To achieve this, techniques are required to analyze the effects of a radiation particle strike on a circuit and evaluate the circuit's resilience to such events. By doing this early in the design flow, significant design effort and resources can be saved. In this chapter, an analytical model was presented for estimating the pulse width of the radiation-induced voltage glitch in combinational circuits. The pulse width of the voltage glitch due to a radiation event is a good measure of radiation robustness of a design. The proposed model efficiently and accurately computes the pulse width of the radiation-induced voltage glitch for any combinational gate. The proposed approach uses a piecewise linear transistor current model and also considers the effect of the ion track establishment constant τ_β of the radiation-induced current pulse, to improve the accuracy of the analysis. Experimental results demonstrate that the proposed model is very fast ($\sim$1,000 times faster than SPICE) and accurate, with a very low pulse width estimation error of 4% compared with SPICE. Thus, the proposed analytical model can therefore be easily incorporated in a design flow to implement radiation tolerant circuits.

References

1. Q. Zhou and K. Mohanram, "Gate sizing to radiation harden combinational logic," *IEEE Transactions on Computer-Aided Design of Integrated Circuits and Systems*, vol. 25, no. 1, pp. 155–166, Jan. 2006.
2. K. Mohanram and N. A. Touba, "Cost-effective approach for reducing soft error failure rate in logic circuits," in *Proc. of the Intl. Test Conf.*, 2003, pp. 893–901.
3. T. Heijmen and A. Nieuwland, "Soft-error rate testing of deep-submicron integrated circuits," in *Proc. of the IEEE European Test Symposium*, 2006, pp. 247–252.
4. K. Mohanram, "Closed-form simulation and robustness models for SEU-tolerant design," in *Proc. of the VLSI Test Symposium*, 2005, pp. 327–333.
5. P. Dahlgren and P. Liden, "A switch-level algorithm for simulation of transients in combinational logic," in *Proc. of the Intl. Symposium on Fault-Tolerant Computing*, June 1995, pp. 207–216.
6. P. E. Dodd and L. W. Massengill, "Basic mechanisms and modeling of single-event upset in digital microelectronics," *IEEE Transactions on Nuclear Science*, vol. 50, no. 3, pp. 583– 602, 2003.
7. Nanoscale integration and modeling (NIMO) group (2007), *ASU Predictive Technology Model [On-line]*, Available: http://www.eas.asu.edu/~ptm.
8. E. Seevinck, F. J. List, and J. Lohstroh, "Static-noise margin analysis of MOS SRAM cells," *IEEE Journal of Solid-State Circuits*, vol. SC-22, no. 5, pp. 748–754, Oct. 1987.
9. T. May and M. Woods, "Alpha-particle-induced soft errors in dynamic memories," *IEEE Transaction on Electron Devices*, vol. ED-26, pp. 2–9, Jan 1979.
10. J. Pickle and J. Blandford, "CMOS RAM cosmic-ray-induced error rate analysis," *IEEE Transactions on Nuclear Science*, vol. NS-29, pp. 3962–3967, 1981.
11. H. Cha, E. M. Rudnick, J. H. Patel, R. K. Iyer, and G. S. Choi, "A gate-level simulation environment for alpha-particle-induced transient faults," *IEEE Transactions on Computers*, vol. 45, no. 11, pp. 1248–1256, 1996.

12. A. Dharchoudhury, S. M. Kang, H. Cha, and J. H. Patel, "Fast timing simulation of transient faults in digital circuits," in *Proc. of the Intl. Conf. on Computer-Aided Design*, 1994, pp. 719–722.
13. P. E. Dodd, F. W. Sexton, and P. S. Winokur, "Three-dimensional simulation of charge collection and multiple-bit upset in Si devices," *IEEE Transactions on Nuclear Science*, vol. 41, pp. 2005–2017, 1994.
14. P. E. Dodd, M. R. Shaneyfelt, and F. W. Sexton, "Charge collection and SEU from angled ion strikes," *IEEE Transactions on Nuclear Science*, vol. 44, pp. 2256–2265, 1997.
15. G. Messenger, "Collection of charge on junction nodes from ion tracks," *IEEE Transactions on Nuclear Science*, vol. 29, no. 6, pp. 2024–2031, 1982.
16. H. Cha, E. M. Rudnick, G. S. Choi, J. H. Patel, and R. K. Iyer, "A fast and accurate gate-level transient fault simulation environment," in *Proc. of the Symposium on Fault-Tolerant Computing*, 1993, pp. 310–319.
17. C. Zhao, X. Bai, and S. Dey, "A scalable soft spot analysis methodology for compound noise effects in nano-meter circuits," in *Proc. of the Design Automation Conf.*, 2004, pp. 894–899.
18. Y. Shih and S. Kang, "Analytic transient solution of general MOS circuit primitives," *IEEE Transactions on Computer-Aided Design of Integrated Circuits and Systems*, vol. 11, no. 6, pp. 719–731, 1992.
19. C. Kashyap, C. Amin, N. Menezes, and E. Chiprout, "A nonlinear cell macromodel for digital applications," in *Proc. of the Intl. Conf. on Computer-Aided Design*, 2007, pp. 678–685.
20. L. Nagel, "Spice: A computer program to simulate computer circuits," in *University of California, Berkeley UCB/ERL Memo M520*, May 1995.

Chapter 3
Analytical Determination of the Radiation-induced Pulse Shape

3.1 Introduction

It was mentioned in last chapter that the circuit hardening approaches [1, 2, 3] often employ selective gate upsizing to reduce the area and delay overhead of the resulting hardened design. These approaches protected only those gates in a circuit which significantly contribute to the soft error failure rate of the circuit. Such gates in the circuit are identified based on three masking factors: logical, electrical, and temporal masking [4, 1]. These masking factors were introduced in Sect. 1.1. All three masking factors reduce the probability of failure due to radiation particle strikes in a combinational circuit. Therefore, for efficient circuit hardening (with low area and delay overheads), it is important to consider the effects of all three masking factors.

Of three masking factors, both logical and temporal masking can be computed without the electrical simulations [4, 1]. However, electrical masking of a gate G in the circuit depends heavily upon the electrical properties of all the gates along any sensitized path from the output of G to any primary output of the circuit. Therefore, efficient and accurate models/simulators for SET events in combinational circuits are required. These simulators should quickly estimate the shape of the voltage glitch at the node where the radiation particle strikes, and then propagate the effect of this voltage glitch to the primary outputs of the circuit. Another reason for the need of the models/simulators for SET events is that when a voltage glitch propagates through the circuit, the pulse width of the voltage glitch can increase, resulting in pulse spreading [5]. With efficient simulators, it will be possible to accurately obtain the glitch width at the primary output of the circuit. This is important for system level circuit hardening approaches [6, 7, 8], which use information about the radiation-induced voltage glitch at the primary output for soft error detection and tolerance mechanisms.

In this chapter, an analytical model is presented, which efficiently estimates the shape of the voltage pulse or glitch that results from a radiation particle strike. The voltage glitch estimated by this analytical model can be propagated to the primary outputs of the circuit using voltage glitch propagation tools such as [9, 10, 11]. The properties of the voltage glitch (such as the magnitude, glitch shape, and width) at the primary outputs can be used to evaluate the radiation robustness of the circuit.

R. Garg and S. P. Khatri, *Analysis and Design of Resilient VLSI Circuits: Mitigating Soft Errors and Process Variations*, DOI 10.1007/978-1-4419-0931-2_3,

On the basis of the result of this analysis, circuit hardening approaches can be implemented to achieve the level of radiation tolerance required.

In the proposed approach for analytical determination of the shape of the radiation-induced voltage glitch, a model for the load current $I_{\text{out}}^{G}(V_{\text{in}}, V_{\text{out}})$ of the output terminal current of the gate G is used. Note that the load current model of the gate is more accurate than the piecewise linear transistor I_{DS} model used in Chap. 2. Again, the model is applicable to any general combinational gate with different loading, and for arbitrary values of collected charge (Q). The effect of the ion track establishment constant (τ_{β}) of the radiation particle induced current pulse is also considered. Experimental results presented in Sect. 3.4 demonstrate that the proposed model for the shape of the radiation-induced voltage glitch is fast and accurate.

The rest of the chapter is organized as follows. Section 3.2 briefly discusses some additional previous work (in addition to the previous work presented in Chap. 2) on modeling of radiation-induced transients in combinational circuits. The model for the shape of the radiation-induced voltage glitch developed in this monograph is described in Sect. 3.3. Experimental results are presented in Sect. 3.4, followed by a chapter summary in Sect. 3.5.

3.2 Related Previous Work

In addition to the previous work already discussed in Sect. 2.2, the authors of [12] presented an iterative approach for soft error rate analysis of combinational circuits (while accounting for electrical masking). As the approach of [12] estimates the effects of a radiation particle strike iteratively, the speedup obtained over SPICE simulations is not high.

A great deal of research has been conducted on circuit-level modeling and simulation for static timing analysis (STA) [13] and static noise analysis (SNA) [14]. These approaches can be extended to estimate the shape of the radiation-induced voltage glitch in combinational circuits. However, the approaches for STA [13] and SNA [14] are iterative, and hence sometimes require a large number of iterations to converge. Thus, the speedup obtained by such iterative approach is not high (the speedup of [13] is 3–70× and [14] is 20× compared with SPICE), and also varies widely depending upon the simulation scenario. In contrast to these iterative approaches, the analytical approach presented in this chapter is *at least* 275× faster compared with SPICE for estimating radiation-induced transients at the output of an inverter.

In [11], the authors developed a general methodology to analyze crosstalk effects in combinational circuits. The authors developed an analytical model for crosstalk excitation. They also developed an analytical model for propagating voltage glitches in combinational circuits. Note that their voltage glitch propagation tool can be used to propagate the radiation-induced voltage glitch estimated by the analytical model presented in this chapter.

3.3 Proposed Analytical Model for the Shape of Radiation-induced Voltage Glitch

Consider four identical inverters as shown in Fig. 3.1 (same as Fig. 2.1a, replicated here for convenience). A radiation particle strike at the node a is modeled by the injection of $i_{seu}(t)$ (described by (1.1)) at node a. As described in Sect. 2.3.1, INV1 (as shown in Fig. 3.1) of different sizes operate quite differently during the radiation-induced transient, and the maximum voltage glitch magnitude (V_{GM}) determines the behavior of their MOSFETs at different times during the transient. The analytical model proposed in this chapter also classifies INV1 (of Fig. 3.1) to be operating in one of four different cases during a radiation-induced transient. The classification is performed in the same manner as described in Sect. 2.3.2. The four cases are briefly described below for completeness.

- Case 1 – $V_{GM} \geq \text{VDD} + V_{diode}$: In this case, with the increasing voltage of node a (V_a), M1 starts conducting in the linear region and enters the saturation region when the V_a becomes more than V_{dsat}^N. M2 starts conducting in the saturation mode once V_a crosses $\text{VDD} + |V_{TP}|$. Eventually when V_a reaches $\text{VDD} + V_{diode}$, the voltage between the source diffusion and the bulk terminal of the PMOS transistor M2 becomes $\geq V_{diode}$. Therefore, the diode between these two terminals gets forward biased and it starts conducting heavily. Thus V_a gets clamped to a value around $\text{VDD} + V_{diode}$. Note that $V_{diode} = 0.6\,\text{V}$.
- Case 2 – $\text{VDD} + |V_{TP}| \leq V_{GM} < \text{VDD} + V_{diode}$: In this case as well, both M1 and M2 conduct similar to Case 1. However, the diode between the diffusion and the bulk terminals of M2 remains off.
- Case 3 – $\text{VDD}/2 \leq V_{GM} < \text{VDD} + |V_{TP}|$: Only M1 conducts in this case. M1 starts conducting in the linear region and when V_a crosses V_{dsat}^N, M1 enters the saturation region. M2 remains off in this case.

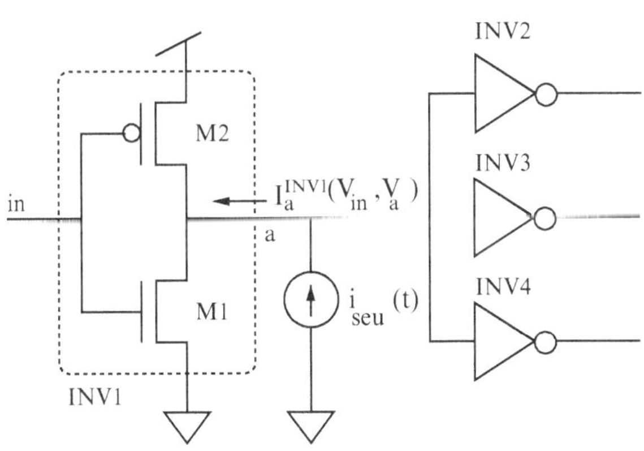

Fig. 3.1 Radiation-induced current injected at the output of inverter INV1

- Case 4 – $V_{GM} < \text{VDD}/2$: This case corresponds to a voltage glitch of magnitude less than VDD/2 and hence the radiation event does not result in a logic flip at the node.

The shape of the radiation-induced voltage glitch is computed differently for different cases, because of the different behavior of M1 and M2 (Fig. 3.1) for these cases (i.e., for Cases 1–3).

An overview of the proposed model is provided in Sect. 3.3.1. Then Sect. 3.3.2 provides details about the proposed method to determine the shape of the radiation-induced voltage glitch.

3.3.1 *Overview of the Proposed Model for Determining the Pulse Shape of the Voltage Glitch*

Similar to the model for the pulse width presented in Chap. 2, the analysis presented in this chapter is also presented for an inverter with its input at VDD and its output at GND. The radiation particle strike results in a positive voltage glitch at the output of the gate. Note that the same analysis (and the same analytical model) for the shape of the radiation-induced voltage glitch can be used for any type of gate (NAND, NOR, etc.), with any logic values applied to its inputs. The handling of NAND, NOR, etc. gates is achieved by constructing an equivalent inverter for the gate. The size of this inverter depends on the given input values of the gate. The applicability of the proposed model to different gates is verified by applying the proposed model to a 2-input NAND gate (for all four input combinations) and 3-input NOR gate (for all eight input combinations). These results are presented in Sect. 3.4. Note that for multiple input gates, the radiation particle strike are not considered at intermediate nodes of the gate, because the worst-case transient occurs when the particle strike occurs at the output node of the gate. Therefore, the estimate of the voltage glitch at the output node due to a particle strike at any intermediate node will not be useful for circuit hardening. Hence, the analysis presented in this chapter is only for radiation strikes at the output node of multi-input gates.

Figure 3.2a (shown at the top left portion of Fig. 3.2) schematically illustrates a voltage glitch that results from a radiation strike at the output node a of INV1. As shown in Fig. 3.2a, the node voltage rises and reaches V_{dsat}^{N} at time T_{sat}^{1}, VDD/2 at time t_1, VDD + $|V_{\text{TP}}|$ at time T_P^1 (for Cases 1 and 2), and then after reaching a maximum value of V_{GM}, the node voltage falls to VDD + $|V_{\text{TP}}|$ at time T_P^2 (for Cases 1 and 2), VDD/2 at time t_2 and finally to V_{dsat}^{N} at the time T_{sat}^{2}. Hence the shape of the voltage glitch of Fig. 3.2a is defined by the node a voltage equations between the time intervals: $(0, T_{\text{sat}}^1)$, $\{(T_{\text{sat}}^1, T_P^1), (T_P^1, T_P^2)$ and $(T_P^2, T_{\text{sat}}^2)\}$ for Cases 1 and 2 or $(T_{\text{sat}}^1, T_{\text{sat}}^2)$ for Case 3, and $(T_{\text{sat}}^2, \infty)$ for all cases. The goal of the proposed approach is to compute the values of all the variables which form the end-points of these time intervals, and also the node voltage equations of node a corresponding to

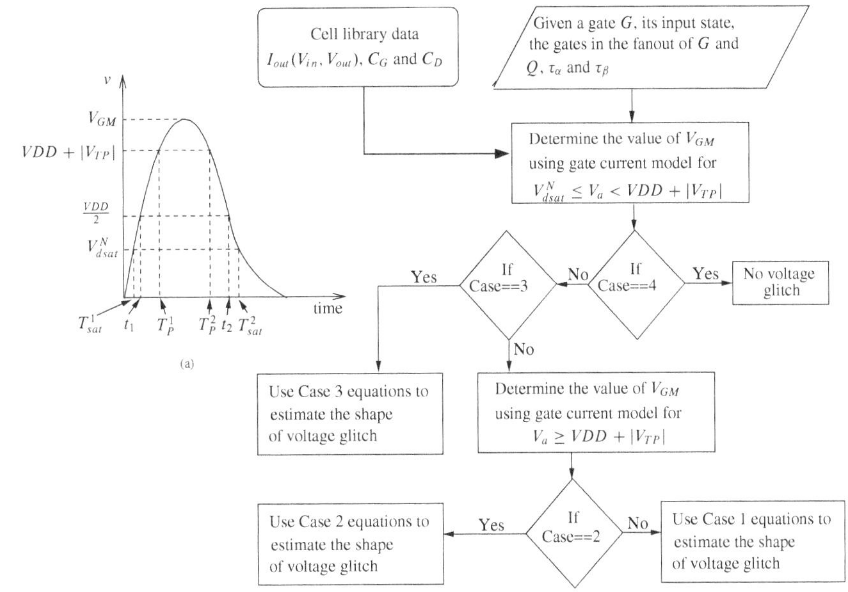

Fig. 3.2 Flowchart of the proposed model for the shape of the radiation-induced voltage glitch

these time durations. The proposed approach can also be used to compute t_1 and t_2 to obtain the width of the voltage glitch (which is $t_2 - t_1$).

All the gates in the library LIB (used in this work) were characterized using the same approach as reported in [13]. For each gate (for all input combinations), the load current of the gate ($I_{\mathrm{out}}(V_{\mathrm{in}}, V_{\mathrm{out}})$) was obtained as a function of its output node voltage, and stored in a look-up table. The input gate capacitance C_{G} (the output node diffusion capacitance C_{D}) was also obtained as a function of the input (output) node voltage and stored in a look-up table. A step size of 0.1 V was used for these look-up table entries. For example, for INV1 of Fig. 3.1, the current through the output terminal a ($I_a(V_{\mathrm{in}}, V_a)$) was obtained for different V_a voltage values at a, when the input node in is at VDD and GND (V_{in} = VDD and V_{in} = GND). Thus, the number of current look-up tables for any gate is equal to 2^n (where n is the number of inputs of a gate). Similarly, C_{D} was also computed based on the input state of the gate. Therefore, for an n-input gate, the total size of the look-up tables for C_{G}, C_{D} and load current I_{out} is $23 \times n$, 17×2^n and 17×2^n, respectively. This characterization step is performed once for each gate in a library and thus it does not affect the runtime of the proposed model. Also, n is typically ≤ 3, hence these lookup tables are quite tractable in practice.

Figure 3.2b shows the flowchart of the algorithm used in the proposed model to compute the shape of the voltage glitch. The input to the model is a gate G (the radiation event is to be simulated at the output node of gate G), its input state, the list of gates which are driven by G, and the values of Q, τ_α and τ_β. The model first computes V_{GM} using the gate current model for $V^N_{\mathrm{dsat}} < V_a <$ VDD $+ |V_{\mathrm{TP}}|$ and then determines the case that is applicable. If $V_{\mathrm{GM}} < \mathrm{VDD}/2$ (i.e., Case 4 applies), then there is no voltage glitch reported. Otherwise if $\mathrm{VDD}/2 < V_{\mathrm{GM}} < \mathrm{VDD} + |V_{\mathrm{TP}}|$ then Case 3 applies and Case 3 equations are used to obtain the shape of the voltage glitch. Otherwise V_{GM} is again computed using the gate current model for $V_a > \mathrm{VDD} + |V_{\mathrm{TP}}|$. Based on this new value of V_{GM}, the operating case of gate G is found (either Case 1 or Case 2) and then the corresponding equations are used to compute the shape of the voltage glitch. The steps of the algorithm used by the model are explained in the following subsections.

3.3.2 *Derivation of the Model for Determining the Shape of the Radiation-induced Voltage Glitch*

As mentioned earlier, the analysis is presented for INV1 (Fig. 3.1) with its input node *in* at VDD and the output node a at GND. A radiation particle strike results in a positive voltage glitch at node a. To ensure that the model for radiation events in combinational circuit elements is manageable, the load current model $I_a^{\mathrm{INV1}}(V_{in}, V_a)$ of INV1 was simplified. Note that in the following analysis $I_a^{\mathrm{INV1}}(V_a)$ is used instead of $I_a^{\mathrm{INV1}}(V_{in}, V_a)$, since the analysis is presented for $V_{in} = \mathrm{VDD}$. With the input terminal of INV1 at VDD, $I_a^{INV1}(V_a)$ can be written as:

$$I_a^{\mathrm{INV1}}(V_a) = \begin{cases} V_a/R_n & V_a < V_{\mathrm{dsat}}^N \\ K_3 + K_4\, V_a & V_{\mathrm{dsat}}^N \le V_a < \mathrm{VDD} + |V_{\mathrm{TP}}| \\ K_5 + K_6\, V_a & \mathrm{VDD} + |V_{\mathrm{TP}}| \le V_a < \mathrm{VDD} + 0.6V. \end{cases}$$

Here, R_n is the linear region resistance of M1 (since M2 is off in this region), which is calculated using the $I_a^{\mathrm{INV1}}(V_a)$ vs. V_a lookup table for V_a values less than V_{dsat}^N. The constants K_3 and K_4 are obtained by using a linear equation for the points $I_a^{\mathrm{INV1}}(V_a)$ vs. V_a from the lookup table for V_a values greater than V_{dsat}^N and less than $\mathrm{VDD} + |V_{\mathrm{TP}}|$. When $V_a > \mathrm{VDD} + |V_{\mathrm{TP}}|$, $I_a^{\mathrm{INV1}}(V_a)$ increases super-linearly with V_a because both M1 and M2 are ON. Thus, the constants K_5 and K_6 are obtained by fitting a least square line to the points $(V_a, I_a^{\mathrm{INV1}}(V_a))$ from the lookup table, for V_a values greater than $VDD + |V_{TP}|$ and less than $\mathrm{VDD} + 0.6V$.

To determine the applicable case, it is first required to find the value of V_{GM}. The method of finding V_{GM} is described next.

3.3.2.1 Voltage Glitch Magnitude V_{GM}

A radiation event can result in a voltage glitch of magnitude greater than VDD/2 flip only if $I_{\mathrm{seu}}^{\max} > I_a^{\mathrm{INV1}}(\mathrm{VDD}/2)$, where $I_{\mathrm{seu}}^{\max}$ is the maximum value of radiation-induced current pulse (1.1). This is a necessary condition which is used to check whether a radiation event will result in a significantly large voltage glitch. The differential equation for the radiation-induced voltage transient at the output of INV1 of Fig. 3.1 is given by:

$$C\frac{\mathrm{d}V_a(t)}{\mathrm{d}t} + I_a^{\mathrm{INV1}}(V_a) = i_{\mathrm{seu}}(t). \tag{3.1}$$

where C is the capacitance[1] at node a. The above equation can be integrated with the initial condition $V_a(t) = 0$ at $t = 0$ to obtain $V_a(t)$. For deep submicron processes, V_{dsat} is much lower than $V_{\mathrm{GS}} - V_T$ due to short channel effects. For the 65 nm PTM [15] model card used in this work, V_{dsat} for both NMOS and PMOS transistors is lower than VDD/2. Therefore, to obtain the V_{GM} value, (3.1) is first integrated from the initial condition and using $I_a^{\mathrm{INV1}} = V_a/R_n$ till V_a reaches the V_{dsat}^N value. Then (3.1) is again integrated using $I_a^{\mathrm{INV1}}(V_a) = K_3 + K_4 V_a$ to obtain the $V_a(t)$ expression. Then, the maximum value V_{GM} attained by this $V_a(t)$ expression is obtained. If $V_{\mathrm{GM}} < \mathrm{VDD} + |V_{\mathrm{TP}}|$ and $V_{\mathrm{GM}} \geq \frac{V_{\mathrm{DD}}}{2}$ then INV1 is in Case 3. Otherwise, INV1 operates in either Case 1 or Case 2[2] if $V_{\mathrm{GM}} \geq V_{\mathrm{DD}} + |V_{\mathrm{TP}}|$. The methodology to decide between Cases 1 and 2 is explained later. Now integrating (3.1) using $I_a^{\mathrm{INV1}}(V_a) = V_a/R_n$ and with the initial condition $V_a(t) = 0$ at $t = 0$, the expression obtained for $V_a(t)$ is:

$$V_a(t) = \frac{I_n}{C}\left(\frac{\mathrm{e}^{-t/\tau_\alpha}}{X} - \frac{\mathrm{e}^{-t/\tau_\beta}}{Y} - Z\mathrm{e}^{-t/R_nC}\right), \tag{3.2}$$

where

$$X = \frac{1}{R_nC} - \frac{1}{\tau_\alpha}, Y = \frac{1}{R_nC} - \frac{1}{\tau_\beta}, I_n = \frac{Q}{\tau_\alpha - \tau_\beta} \& Z = \frac{1}{X} - \frac{1}{Y}.$$

To obtain the time T_{sat}^1 when $V_a(t)$ reaches the V_{dsat}^N value from (3.2), linearly expand (3.2) around the initial guess T_{sat}^{1a}. The expression for T_{sat}^1 thus obtained is:

$$T_{\mathrm{sat}}^1 = T_{\mathrm{sat}}^{1a} + \frac{V_{\mathrm{dsat}}^N - \frac{I_n}{C}\left(\frac{\mathrm{e}^{-T_{\mathrm{sat}}^{1a}/\tau_\alpha}}{X} - \frac{\mathrm{e}^{-T_{\mathrm{sat}}^{1a}/\tau_\beta}}{Y} - Z\mathrm{e}^{-T_{\mathrm{sat}}^{1a}/R_nC}\right)}{\frac{I_n}{C}\left(-\frac{\mathrm{e}^{-T_{\mathrm{sat}}^{1a}/\tau_\alpha}}{\tau_\alpha X} + \frac{\mathrm{e}^{-T_{\mathrm{sat}}^{1a}/\tau_\beta}}{\tau_\beta Y} + \frac{Z}{R_nC}\mathrm{e}^{-T_{\mathrm{sat}}^{1a}/R_nC}\right)}. \tag{3.3}$$

To obtain the initial guess T_{sat}^{1a}, approximate the rising part of the radiation-induced current by a line between the origin and the point where $i_{\mathrm{seu}}(t)$ of (1.1) reaches its maximum value $I_{\mathrm{seu}}^{\max}$. The radiation-induced current $i_{\mathrm{seu}}(t)$ reaches $I_{\mathrm{seu}}^{\max}$ at T_{seu}^{max}. Then substitute this approximated radiation current in the RHS of (3.1)

---

[1] The value of C is obtained by the addition of the average value of $n \times C_{\mathrm{G}}$, C_{D} and the capacitance of interconnect over the operating voltage range. Here, n is the fanout factor.

[2] In Cases 1 and 2, both M1 and M2 conduct and hence the INV1 load current model $K_5 + K_6 V_a$ is used to obtain accurate value of V_{GM}. This new value of V_{GM} is used to resolve between Cases 1 and 2.

and integrate it from the initial condition $V_a(t) = 0$ at $t = 0$ to $V_a(t) = V_{\text{dsat}}^N$ at $t = T_{\text{sat}}^{1a}$ using $I_a^{INV1}(V_a) = V_a/R_n$. After this, solve for T_{sat}^{1a} by performing a quadratic expansion of the resulting equation around the origin. The expression for T_{sat}^{1a} is:

$$T_{\text{sat}}^{1a} = \sqrt{\frac{2V_{\text{dsat}}^N\, C\, T_{\text{seu}}^{\max}}{I_{\text{seu}}^{\max}}}, \tag{3.4}$$

where

$$T_{\text{seu}}^{\max} = \frac{\tau_\alpha \tau_\beta}{\tau_\alpha - \tau_\beta} \log \frac{\tau_\alpha}{\tau_\beta} \quad \text{and} \quad I_{\text{seu}}^{\max} = i_{\text{seu}}\left(T_{\text{seu}}^{\max}\right).$$

So far T_{sat}^1 (the time when $V_a(t)$ reaches V_{dsat}^N, or the time when M1 enters the saturation mode) is known. Now, again integrate (3.1) with the initial condition $V_a(t) = V_{\text{dsat}}^N$ at $t = T_{\text{sat}}^1$, and using $I_a^{\text{INV1}}(V_a) = K_3 + K_4 V_a$. The resulting expression for $V_a(t)$ is:

$$V_a(t) = \frac{I_n}{C}\left(\frac{\mathrm{e}^{-t/\tau_\alpha}}{X'} - \frac{\mathrm{e}^{-t/\tau_\beta}}{Y'}\right) - \frac{K_3}{K_4} + Z'\mathrm{e}^{-K_4 t/C} \tag{3.5}$$

where,

$$X' = \frac{K_4}{C} - \frac{1}{\tau_\alpha},\ Y' = \frac{K_4}{C} - \frac{1}{\tau_\beta} \quad \text{and}$$

$$Z' = V_{\text{dsat}}^N \mathrm{e}^{K_4 T_{\text{sat}}^1/C} - \frac{I_n}{C}\mathrm{e}^{K_4 T_{\text{sat}}^1/C}\left(\frac{\mathrm{e}^{-T_{\text{sat}}^1/\tau_\alpha}}{X'} - \frac{\mathrm{e}^{-T_{\text{sat}}^1/\tau_\beta}}{Y'}\right) + \frac{K_3}{K_4}\mathrm{e}^{K_4 T_{\text{sat}}^1/C}.$$

To calculate the value of V_{GM}, first differentiate (3.5) and equate $\mathrm{d}V_a(t)/\mathrm{d}t$ to zero and solve for $T_{V_{\text{GM}}}$ (the time at which $V_a(t)$ reaches its maximum value). Since the equation $\mathrm{d}V_a(t)/\mathrm{d}t = 0$ is also a transcendental equation, hence linearly expand $\mathrm{d}V_a(t)/\mathrm{d}t = 0$ around $T_{V_{\text{GM}}}^a$ and then solve for $T_{V_{\text{GM}}}$. The expression obtained for $T_{V_{\text{GM}}}$ is:

$$T_{V_{\text{GM}}} = T_{V_{\text{GM}}}^a + \frac{\frac{\mathrm{e}^{-T_{V_{\text{GM}}}^a/\tau_\alpha}}{\tau_\alpha X'} - \frac{\mathrm{e}^{-T_{V_{\text{GM}}}^a/\tau_\beta}}{\tau_\beta Y'} + \frac{K_4 Z'}{C}\mathrm{e}^{-K_4 T_{V_{\text{GM}}}^a/C}}{\frac{\mathrm{e}^{-T_{V_{\text{GM}}}^a/\tau_\alpha}}{\tau_\alpha^2 X'} - \frac{\mathrm{e}^{-T_{V_{\text{GM}}}^a/\tau_\beta}}{\tau_\beta^2 Y'} + \frac{K_4^2 Z'}{C^2}\mathrm{e}^{-K_4 T_{V_{\text{GM}}}^a/C}}. \tag{3.6}$$

Now, calculate V_{GM} by substituting $T_{V_{\text{GM}}}$ obtained from (3.6), into (3.5). If $V_{\text{GM}} < \text{VDD}/2$ then Case 4 applies and the radiation event does not flip the logic level of the affected node. If $\text{VDD}/2 \leq V_{\text{GM}} < \text{VDD} + |V_{\text{TP}}|$, then Case 3 is applicable. Otherwise, either Case 1 or Case 2 is applicable. Before describing the methodology to decide between Case 1 and Case 2, the method to obtain the value of $T_{V_{\text{GM}}}^a$ is first discussed.

Note that the output node voltage of INV1 (i.e., $V_a(t)$ of (3.5)) always attains its maximum value after $T_{\rm seu}^{\rm max}$ (the time $i_{\rm seu}(t)$ of (1.1) reaches its maximum value I_{seu}^{max}). Therefore, integrate (3.1) using a linear model ($i_{\rm seu}^m(t)$) for the radiation-induced current for time $t > T_{\rm seu}^{\rm max}$ and with the initial condition $V_a(t) = V_a^{\rm sm}$ at $t = T_{\rm seu}^{\rm max}$ (obtained from (3.5)). The radiation-induced linear current model $i_{\rm seu}^m(t)$ has one of its end-points $I_{\rm seu}^{\rm max}$ at a time value of $T_{\rm seu}^{\rm max}$. The other end-point has its current value of 0, and its time value t^* is obtained by equating the charge deposited by the actual radiation-induced current $i_{\rm seu}(t)$ from time $T_{\rm seu}^{\rm max}$ to ∞ and the charge deposited by the linearized radiation-induced current equation. Hence the expression for the radiation-induced linear current model is:

$$i_{\rm seu}^m(t) = I_{\rm seu}^{\rm max}\left(1 - \frac{t - T_{\rm seu}^{\rm max}}{t^* - T_{\rm seu}^{\rm max}}\right) = P + Mt. \tag{3.7}$$

Now substitute $i_{\rm seu}^m(t)$ for $i_{\rm seu}(t)$ in (3.1), use $I_a^{\rm INV1}(V_a) = K_3 + K_4\, V_a$ and then integrate. After this, differentiate the resulting equation for $V_a(t)$ and equate $\mathrm{d}V_a(t)/\mathrm{d}t$ to zero and solve for $T_{V_{\rm GM}}^a$.

Deciding Between Case 1 and Case 2. Before deciding whether INV1 is operating in Case 1 or Case 2, it is first required to compute the time t_1 when $V_a(t)$ reaches VDD/2. Then T_P^1 (the time when $V_a(t)$ reaches VDD + $|V_{TP}|$) is computed using t_1. After this, integrate (3.1) using the initial condition $V_a(t) = \mathrm{VDD} + |V_{\rm TP}|$ at $t = T_P^1$ and $I_a^{\rm INV1}(V_a) = K_5 + K_6\, V_a$, to obtain the expression for $V_a(t)$. Then this expression of $V_a(t)$ will be used to decide between Cases 1 and 2 using the $V_{\rm GM}$ value. As shown in the flowchart of the algorithm of the proposed approach in Fig. 3.2, the method to compute t_1 is identical for cases 1, 2, or 3. To obtain the expression for t_1, substitute $t = t_1$ and $V_a(t_1)$ = VDD/2 in (3.5) and then solve for t_1 after expanding it linearly around the point t_1^a (which is an initial guess for t_1). Here $t_1^a = T_{\rm sat}^1 \mathrm{VDD}/(2V_{\rm dsat}^N)$. The expression for t_1 is therefore:

$$t_1 = t_1^a + \frac{\frac{e^{-t_1^a/\tau_\alpha}}{X'} - \frac{e^{-t_1^a/\tau_\beta}}{Y'} + \frac{C}{I_n}\left(Z' e^{-K_4 t_1^a/C} - \frac{K_3}{K_4} - \frac{VDD}{2}\right)}{\frac{e^{-t_1^a/\tau_\alpha}}{X'\tau_\alpha} - \frac{e^{-t_1^a/\tau_\beta}}{Y'\tau_\beta} + \frac{K_4 Z'}{I_n} e^{-K_4 t_1^a/C}}. \tag{3.8}$$

Then compute the time $t = T_P^1$ when $V_a(t)$ reaches VDD + $|V_{\rm TP}|$, since the load current model of INV1 changes at this time instant. To obtain T_P^1, repeat the same steps followed for the derivation of the t_1 expression with the condition $V_a(t) = \mathrm{VDD} + |V_{TP}|$ at $t = T_P^1$ in (3.5), and with the initial guess $T_P^{1a} = t_1 + (\mathrm{VDD} + |V_{TP}| - V_{\rm dsat}^N)/(\mathrm{VDD}/2 - V_{\rm dsat}^N)$. The expression for T_P^1 is therefore similar to (3.8) with t_1^a replaced by T_P^{1a}, t_1 by T_P^1 and $VDD/2$ by VDD + $|V_{\rm TP}|$.

Now integrate (3.1) with the initial condition $V_a(t) = \mathrm{VDD} + |V_{\rm TP}|$ at $t = T_P^1$, and using $I_a^{\rm INV1}(V_a) = K_5 + K_6\, V_a$. The resulting expression for $V_a(t)$ is:

$$V_a(t) = \frac{I_n}{C}\left(\frac{\mathrm{e}^{-t/\tau_\alpha}}{X''} - \frac{\mathrm{e}^{-t/\tau_\beta}}{Y''}\right) - \frac{K_5}{K_6} + Z''\mathrm{e}^{-K_6 t/C}, \tag{3.9}$$

where

$$X'' = \frac{K_6}{C} - \frac{1}{\tau_\alpha}, \quad Y'' = \frac{K_6}{C} - \frac{1}{\tau_\beta},$$

$$Z'' = V_{\mathrm{dsat}}^N \mathrm{e}^{K_6 T_P^1/C} - \frac{I_n}{C}\mathrm{e}^{K_6 T_P^1/C}\left(\frac{\mathrm{e}^{-T_P^1/\tau_\alpha}}{X''} - \frac{\mathrm{e}^{-T_P^1/\tau_\beta}}{Y''}\right) + \frac{K_5}{K_6}\mathrm{e}^{K_6 T_P^1/C}.$$

To calculate the value of the maximum value of $V_a(t)$ of (3.9) (i.e., V_{GM}, the maximum glitch magnitude for Case 1 or 2), repeat the same steps is followed while calculating the maximum value of $V_a(t)$ of (3.5). After obtaining the value of V_{GM}, it can be decided whether INV1 is operating in Case 1 or 2. Note that by using this method, V_{GM} can be evaluated to be greater than VDD $+ 0.6V$, because the diode is not modeled in (3.1). Therefore, if $V_{\mathrm{GM}} >$ VDD $+ 0.6V$ then V_{GM} is set to VDD $+ 0.6V$.

So far, the expression for V_{GM} is known, which can be used to determine the operating case of INV1. Also, expressions were derived for T_{sat}^1, t_1, T_P^1 and the INV1 output node voltage equations for different time durations (3.2, 3.5, 3.9).

3.3.2.2 Derivation of the Expressions for Case 3

The derivation of the expressions for the shape of the voltage glitch in Case 3 is as follows. First, derive the expression for t_2 i.e., the time when $V_a(t)$ falls to the VDD$/2$ value. Note that in this case, only M1 of Fig. 3.1a conducts because the magnitude of the glitch voltage is less than VDD$+|V_{\mathrm{TP}}|$. Therefore, (3.5) describes the voltage of node a for all times t such that $T_{\mathrm{sat}}^1 \le t \le T_{\mathrm{sat}}^2$. The expression for t_2 can be obtained in similar manner as t_1, with the substitution of $t = t_2$ and $V_a(t_2) =$ VDD$/2$ in (3.5) and with the initial guess point t_2^a. It was observed that the time when $i_{\mathrm{seu}}(t)$ falls to $I_{\mathrm{DS}}^{\mathrm{VDD}/2}$ after reaching $I_{\mathrm{seu}}^{\max}$ can be used as an initial guess (t_2^a) for t_2 since the node voltage at that time will be close to VDD$/2$. The contribution of the e^{-t/τ_β} term of $i_{\mathrm{seu}}(t)$ was ignored when calculating t_2^a. This is reasonable since τ_α is usually 3–4 times of τ_β and therefore $\mathrm{e}^{-t/\tau_\beta}$ approaches 0 much faster than the $\mathrm{e}^{-t/\tau_\alpha}$ term. Thus the value of $\mathrm{e}^{-t/\tau_\beta}$ around t_2^a (which is greater than $T_{\mathrm{seu}}^{\max}$) will be approximately equal to 0. The expression for t_2^a is $-\tau_\alpha \log I_{\mathrm{DS}}^{\mathrm{VDD}/2}/I_n$.

Now, again substitute $t = T_{\mathrm{sat}}^2$ and $V_a(T_{\mathrm{sat}}^2) = V_{\mathrm{dsat}}^N$ in (3.5) and solve for T_{sat}^2 in a similar manner as solved for t_1 (3.8) using the initial guess T_{sat}^{2a}. The expression for T_{sat}^{2a} is $t_2 + V_{\mathrm{dsat}}^N - 0.5 \times \mathrm{VDD}/(\mathrm{d}V_a(t)/\mathrm{d}t|_{t=t_2})$.

To obtain the node a voltage equation for $t > T_{\mathrm{sat}}^2$, integrate (3.1) with the initial condition $V_a(T_{\mathrm{sat}}^2) = V_{\mathrm{dsat}}^N$ and using $I_a^{\mathrm{INV1}}(V_a) = V_a/R_n$. The expression thus obtained is:

$$V_a(t) = \frac{I_n}{C}\left(\frac{\mathrm{e}^{-t/\tau_\alpha}}{X} - \frac{\mathrm{e}^{-t/\tau_\beta}}{Y} - A_p \mathrm{e}^{(T_{\mathrm{sat}}-t)/R_n C}\right), \tag{3.10}$$

where

$$A_p = V_{\text{dsat}}^N - \frac{I_n}{C}\left(\frac{e^{-T_{\text{sat}}^2/\tau_\alpha}}{X} - \frac{e^{-T_{\text{sat}}^2/\tau_\beta}}{Y}\right).$$

Now the analytical expression of the radiation-induced voltage glitch for Case 3 is complete. The voltage glitch is described by a set of three equations (3.2, 3.5, and 3.10), as summarized below:

$$V_a(t) = \begin{cases} \text{Equation (3.2)} & t < T_{sat}^1 \\ \text{Equation (3.5)} & T_{sat}^1 \leq t \leq T_{sat}^2 \\ \text{Equation (3.10)} & t > T_{sat}^2. \end{cases}$$

3.3.2.3 Derivation of the Expressions for Case 2

In this case, the magnitude of the voltage glitch V_{GM} is between VDD+$|V_{\text{TP}}|$ and VDD+0.6V. Therefore, both M1 and M2 of INV1 conduct for a time t such that $T_P^1 \leq t \leq T_P^2$. Hence node a's voltage is described by (3.9) (this equation was used to calculate the V_{GM} value for Cases 1 and 2). To obtain the value of T_P^2, substitute $V_a(T_P^2) = \text{VDD} + |V_{\text{TP}}|$ for $t = T_P^2$ in (3.9) and then solve for T_P^2 by using T_P^{2a} as the initial guess. The resulting expression for T_P^2 is:

$$T_P^2 = T_P^{2a} + \frac{\frac{e^{-T_P^{2a}/\tau_\alpha}}{X''} - \frac{e^{-T_P^{2a}/\tau_\beta}}{Y''} + \frac{C}{I_n}\left(Z'' e^{-K_6 T_P^{2a}/C} - \frac{K_5}{K_6} - (\text{VDD} + |V_{\text{TP}}|)\right)}{\frac{e^{-T_P^{2a}/\tau_\alpha}}{X''\tau_\alpha} - \frac{e^{-T_P^{2a}/\tau_\beta}}{Y''\tau_\beta} + \frac{K_6 Z''}{I_n} e^{-K_6 T_P^{2a}/C}}. \tag{3.11}$$

The value of T_P^{2a} is obtained using the following observation. When $i_{\text{seu}}(t)$ becomes equal to the drain to source current (I_{DS}) of M1 of Fig. 3.1a, then at that instant, the I_{DS} of M2 is approximately equal to 0 and the voltage at node a is VDD $+|V_{\text{TP}}|$. Thus, the value of T_P^{2a} is obtained by solving $I_a^{\text{INV1}}(\text{VDD} + |V_{TP}|) = i_{\text{seu}}(T_P^{2a})$ (since at this instant I_{DS} of M2 is zero therefore I_{DS} of M1 is equal to $I_a^{\text{INV1}}(\text{VDD} + |V_{\text{TP}}|)$). In this derivation, the contribution of the e^{-t/τ_β} term of $i_{\text{seu}}(t)$ is ignored for the reason explained in Sect. 3.3.2.2. The expression for T_P^{2a} is $-\tau_\alpha \log(I_a^{\text{INV1}}(\text{VDD} + |V_{\text{TP}}|)/I_n)$.

Now calculate the voltage equation of node a for time duration $T_P^2 \leq t \leq T_{\text{sat}}^2$. For this, integrate (3.1) with the initial condition $V_a(t) = \text{VDD} + |V_{\text{TP}}|$ at $t = T_P^2$, and using $I_a^{\text{INV1}}(V_a) = K_3 + K_4 V_a$. The resulting expression for $V_a(t)$ thus obtained is:

$$V_a(t) = \frac{I_n}{C}\left(\frac{e^{-t/\tau_\alpha}}{X'} - \frac{e^{-t/\tau_\beta}}{Y'}\right) - \frac{K_3}{K_4} + Z^* e^{-K_4 t/C}, \tag{3.12}$$

where

$$Z^* = (\mathrm{VDD} + |V_{\mathrm{TP}}|)\mathrm{e}^{K_4 T^1_{\mathrm{sat}}/C} - \frac{I_n}{C}\mathrm{e}^{K_4 T^2_P/C}\left(\frac{\mathrm{e}^{-T^2_P/\tau_\alpha}}{X'} - \frac{\mathrm{e}^{-T^2_P/\tau_\beta}}{Y'}\right) + \frac{K_3}{K_4}\mathrm{e}^{K_4 T^2_P/C}.$$

Using (3.12), the values of t_2 and T^2_{sat} can be obtained for Case 2 in the same manner as t_2 and T^2_{sat} were derived for Case 3. After finding the values for t_2 and T^2_{sat}, the voltage equation of node a, for $t > T^2_{\mathrm{sat}}$ is same as (3.10) (with the values of t_2 and T^2_{sat} calculated for this case). Now all variables for this case have been derived. The equation for the radiation-induced voltage glitch at node a is as shown below:

$$V_a(t) = \begin{cases} \text{Equation (3.2)} & t < T^1_{\mathrm{sat}} \\ \text{Equation (3.5)} & T^1_{\mathrm{sat}} \le t < T^1_P \\ \text{Equation (3.9)} & T^1_P \le t < T^2_P \\ \text{Equation (3.12)} & T^2_P \le t \le T^2_{\mathrm{sat}} \\ \text{Equation (3.10)} & t > T^2_{\mathrm{sat}}. \end{cases}$$

3.3.2.4 Derivation of the Expressions for Case 1

In this case, both M1 and M2 of Fig. 3.1a conduct, similar to Case 2. However, when the voltage at node a reaches a value VDD $+ 0.6V$, the diffusion diode between node a and the bulk terminal of M2 gets forward biased and starts conducting heavily. Thus $V_a(t)$ gets clamped to a value around VDD $+ 0.6V$. Therefore, all expressions derived for Case 2 are also applicable to this case, with a slight modification to incorporate the effect of the diode clamping action. In this case, when (3.9) computes a value greater than VDD $+ 0.6V$ for any time t then the voltage of node a is set to VDD $+ 0.6V$. Thus, the resulting equations for the voltage glitch for this case are:

$$V_a(t) = \begin{cases} \text{Equation (3.2)} & t < T^1_{\mathrm{sat}} \\ \text{Equation (3.5)} & T^1_{\mathrm{sat}} \le t < T^1_P \\ min(\text{Equation (3.9)}, \mathrm{VDD} + 0.6V) & T^1_P \le t < T^2_P \\ \text{Equation (3.12)} & T^2_P \le t \le T^2_{\mathrm{sat}} \\ \text{Equation (3.10)} & t > T^2_{\mathrm{sat}} \end{cases}$$

The equations for the radiation-induced voltage glitch derived in this Section (for Cases 1–3) determine the shape of the glitch. Note that τ_β was *not ignored* in the derivation of the voltage glitch equations and in the calculation of all time variables of the proposed model such as T^1_{sat}, t_1, T^1_P, etc. Sometimes, the contribution of the e^{-t/τ_β} term of $i_{\mathrm{seu}}(t)$ was ignored, but this was done only during the *calculation of the initial guess* for these time variables.

3.4 Experimental Results

The accuracy of the proposed model for determining the shape of the radiation-induced voltage glitch was compared with SPICE [16]. The proposed model was implemented in *perl* and was determined to be 275× faster than SPICE for the estimation of the radiation-induced voltage glitch at the output of an inverter. For other gates such as NAND, NOR, etc., SPICE takes more time to simulate a radiation particle strike, because of the larger number of transistors in these gates, compared with an inverter. However, the runtime of the proposed approach does not change significantly with different gate types, because of the utilization of a load current model for all gates. Therefore, the speedup of the model proposed in this chapter, compared with SPICE simulation, will be higher for NAND, NOR, and complex gates [3].

A standard cell library *LIB* was implemented using a 65 nm PTM [15] model, card with VDD = $1V$. The library *LIB* contains INV, NAND, and NOR gates of five different sizes (1× to 5×) with different numbers of inputs. As mentioned in Sect. 3.3.1, the look-up tables for the load current model of the gate, the input gate capacitance C_G and the output node diffusion capacitance C_D (for all input combinations) were obtained for all the gates in *LIB*. The method to obtain the load current, and the C_G and C_D look-up tables was explained in Sect. 3.3.1.

To validate the applicability of the proposed model to different types of gates, radiation particle strikes were simulated at the output of INVs, 2-input NANDs, and 3-input NORs, using the proposed model. For each gate type, five different sizes (1× to 5×) were considered, with all possible input states. The applicability of the model to different scenarios was also validated by loading the gates with different loads, and by varying the values Q, τ_α and τ_β. All gates were loaded with 1 and 3 inverters of the same size as the equivalent inverter of G. The radiation particle strikes were simulated corresponding to $Q = 150\,\text{fC}$, $\tau_\alpha = 150\,\text{ps}$, and $\tau_\beta = 50\,\text{ps}$ and $Q = 100\,\text{fC}$, $\tau_\alpha = 200\,\text{ps}$, and $\tau_\beta = 50\,\text{ps}$.

The radiation-induced voltage glitches obtained using the proposed model and SPICE are shown in Fig. 3.3 for the INV, NAND2, and NOR3 gates, with different scenarios (as mentioned in the figure). Figure 3.3 also reports the operating case for the gate along with the gate size and the input state. From Fig. 3.3, observe that the voltage glitch waveforms obtained using the proposed model *match very closely* with the voltage glitch obtained from SPICE. Note that INV, NAND2, and NOR3 of different sizes with all possible input states and with different radiation-induced current pulses were simulated. However, for brevity only a few representative waveforms are shown in Fig. 3.3. The waveforms shown in Fig. 3.3 were chosen to demonstrate the applicability of the proposed model to different scenarios. Figure 3.3b corresponds to a Case 3 scenario in which a 4× INV has its input at GND value and is driving 3–4× INVs. In this case, the voltage glitch predicted by the model deviates from SPICE when the affected node voltage drops to 0.2 V. This

[3] For a 2-input NAND gate, the proposed model is 330× faster than SPICE simulations.

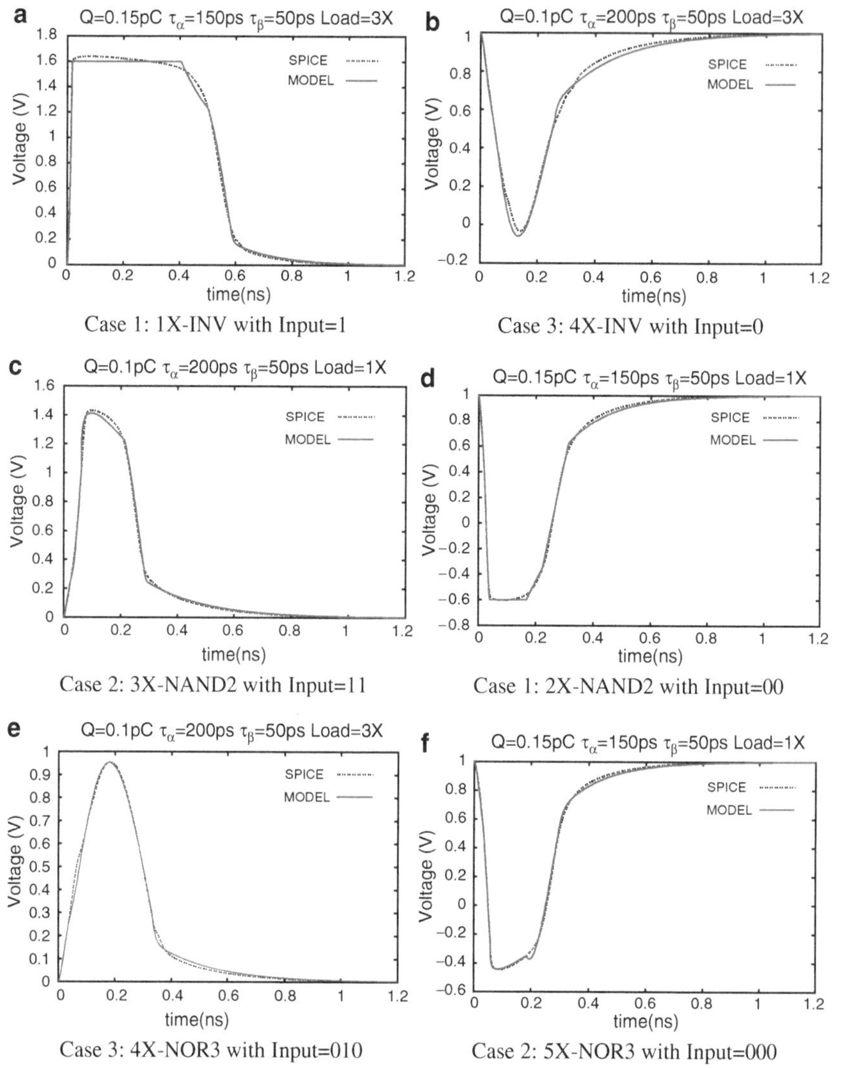

Fig. 3.3 Radiation-induced voltage glitches obtained using the proposed model and SPICE for different gates

is due to the Miller feedback from the switching of the output of the loading inverters (3–4× INVs) to the node affected by the radiation strike. The effect of the Miller feedback is more dominant for the gates operating in Case 3, than for Case 1 and 2. This is because in Case 3, the effect of a radiation particle strike is lower than in Case 1 or 2, and hence the Miller feedback has a significant impact on the voltage glitch. Slight mismatches can also be observed in some of the voltage glitch waveforms of Fig. 3.3. This is due to the modeling error, which is introduced by gate characterization (which is performed with a voltage step of 0.1 V).

Table 3.1 RMSP error of the proposed model for 3× gates and $Q = 150\,\text{fC}$, $\tau_\alpha = 150\,\text{ps}$, and $\tau_\beta = 50\,\text{ps}$

		Input state								
Load	Gate	0	1	2	3	4	5	6	7	Avg. RMSP err.
1	INV	2.86	2.62							2.74
1	NAND2	3.75	3.05	3.3	4.0					3.52
1	NOR3	3.45	2.43	7.06	3.65	10.85	5.38	7.40	8.76	6.12
3	INV	3.46	4.94							4.2
3	NAND2	3.72	3.36	3.57	5.89					4.13
3	NOR3	3.29	4.24	5.13	4.51	9.40	5.72	5.64	10.41	6.04
AVG										5.06

The performance of the model was quantified by calculating the root-mean-square-percentage (*rmsp*) error of the voltage glitches obtained using the model, compared with the glitch waveforms obtained using SPICE. Note that the *rmsp* error was used to compare the accuracy of the proposed model with SPICE because the goal of the proposed model is to accurately estimate the radiation-induced voltages transient waveform (voltage glitch). This voltage glitch then can be propagated to the primary outputs of the circuit using voltage glitch propagation tools such as [9, 10, 11] to evaluate the radiation robustness of the circuit. The *rmsp* error was computed over a time period for which the affected node voltage value is greater (lesser) than V_{TN} (VDD $-$ $|V_{TP}|$) for a positive (negative) glitch. Table 3.1 reports the *rmsp* error of the model for 3× gates and with a radiation particle strike with $Q = 150\,\text{fC}$, $\tau_\alpha = 150\,\text{ps}$, and $\tau_\beta = 50\,\text{ps}$, for all possible input states. Column 1 reports the number of inverters driven by the gate reported in Column 2. Note that the loading inverters are of the same size as the equivalent inverter of the corresponding gate. Columns 3–10 report the *rmsp* error of the voltage glitch estimated by the model, compared with SPICE, for all possible input states. Column 11 reports the average *rmsp* error for a 3× gate, averaged over its all possible states. A blank entry in Table 3.1 indicates that the input state of the corresponding column is not applicable to the corresponding gate. Observe from Table 3.1 that the proposed model is able to predict the radiation-induced voltage glitch for 3× gates with a very small *rmsp* error of 5.06% (as reported by the last row of Table 3.1) averaged over all gates for all input states. Similar results were obtained for $Q = 100\,\text{fC}$, $\tau_\alpha = 200\,\text{ps}$ and $\tau_\beta = 50\,\text{ps}$.

Table 3.2 reports the *rmsp* error of the proposed model, for different gate sizes (from 1× to 5×) with $Q = 150 fC$, $\tau_\alpha = 150\,\text{ps}$ and $\tau_\beta = 50\,\text{ps}$, averaged over all possible input states for the gate. Table 3.2 shows that the proposed model to estimate the shape of the radiation-induced voltage glitch is very accurate and the average *rmsp* error is 4.45% averaged over all simulated scenarios (different gate types, gate loading, and gate sizes). Also, the proposed approach is *at least* 275× faster than SPICE simulations. Note that the best known previous analytical approach to predict the radiation-induced voltage glitch is reported in [17] to be just 100× faster than SPICE. Also in [17], the authors report that their approach sometimes yields a 15% error in the radiation-induced glitch, compared with SPICE.

Table 3.2 RMSP error of the proposed model for different gates sizes and $Q = 150\,\text{fC}$, $\tau_\alpha = 150\,\text{ps}$, and $\tau_\beta = 50\,\text{ps}$

		Gate Size					
Load	Gate	1×	2×	3×	4×	5×	Avg. RMSP err.
1	INV	2.72	2.66	2.74	3.08	3.49	2.94
1	NAND2	3.45	3.27	3.52	3.93	3.80	3.6
1	NOR3	4.43	4.76	6.04	6.96	6.02	5.64
3	INV	3.66	3.95	4.20	4.61	5.15	4.3
3	NAND2	3.81	3.83	4.14	4.69	4.53	4.2
3	NOR3	4.77	4.99	6.12	6.98	7.12	6.00
AVG							4.45

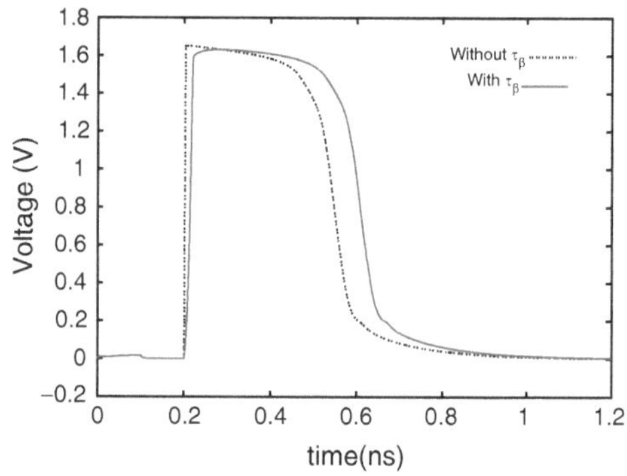

Fig. 3.4 Radiation-induced voltage glitch at 2X-INV1

Moreover, the authors ignore the effect of the ion track establishment constant (τ_β), by setting it to zero for both their model as well as for their SPICE simulations. To evaluate the impact of ignoring τ_β on the radiation-induced voltage glitch, radiation particle strikes were simulated in SPICE (with and without the inclusion of τ_β) at the output of inverters of different sizes (1× - 5×). These simulations were performed for two different radiation strike parameter values ($Q = 150$ fC, $\tau_\alpha = 150\,\text{ps}$, and $\tau_\beta = 50\,\text{ps}$), as well as ($Q = 100$ fC, $\tau_\alpha = 200\,\text{ps}$ and $\tau_\beta = 50\,\text{ps}$) and for different loads on the inverters. For $Q = 150\,\text{fC}$, $\tau_\alpha = 150\,\text{ps}$, and $\tau_\beta = 50\,\text{ps}$ ($Q = 100\,\text{fC}$, $\tau_\alpha = 200\,\text{ps}$, and $\tau_\beta = 50\,\text{ps}$), it was found that ignoring τ_β results in an *underestimation of the pulse width of the voltage glitch by 10% (8%)*. The voltage waveforms at the output of a 2× inverter under a radiation particle strike with and without the inclusion of the τ_β term (for $Q = 150\text{fC}$, $\tau_\alpha = 150\,\text{ps}$, and $\tau_\beta = 50\,\text{ps}$) are shown in Fig. 3.4. The *rmsp* error of the voltage glitch without τ_β (shown in Fig. 3.4) is 40%, which is much higher than the error of the proposed approach. Thus, for an accurate analysis, it is crucial to include the contribution of τ_β. As mentioned earlier, the authors of [17] ignore τ_β and therefore, the error of their approach can be much higher than reported in [17], when compared with the shape of the radiation-induced voltage glitch obtained while considering the contributions of τ_β. The analytical model presented in this chapter is the first to model the effect of both τ_α and τ_β, for the estimation of the shape of the radiation-induced voltage glitch. Thus, the proposed approach is more accurate than the best known previous approach [17].

3.5 Chapter Summary

In this chapter, an analytical model for the determination of the shape of radiation-induced voltage glitches in combinational circuits was presented. The radiation-induced voltage glitch at an internal node of a circuit can be propagated to the primary outputs of the circuit (using existing tools [9, 10, 11]) to account for the effects of electrical masking. This enables an accurate and quick evaluation of the radiation robustness of a circuit. Experimental results demonstrate that the proposed model is very accurate, with a very low root mean square percentage error in the estimation of the shape of the voltage glitch (of 4.5%) compared with SPICE. The model presented in this chapter gains its accuracy by using a piecewise-linear model for the load current of the gate, and by considering the effect of τ_β of the radiation-induced current pulse. The analytical model is very fast (275× faster than SPICE) and accurate, and can therefore be easily incorporated in a design flow to implement radiation tolerant circuits. The next chapter presents a model for the dynamic stability of a SRAM cell during a radiation particle strike.

References

1. Q. Zhou and K. Mohanram, "Gate sizing to radiation harden combinational logic," *IEEE Transactions on Computer-Aided Design of Integrated Circuits and Systems*, vol. 25, no. 1, pp. 155–166, Jan. 2006.
2. T. Heijmen and A. Nieuwland, "Soft-error rate testing of deep-submicron integrated circuits," in *Proc. of the IEEE European Test Symposium*, 2006, pp. 247–252.
3. K. Mohanram and N. A. Touba, "Cost-effective approach for reducing soft error failure rate in logic circuits," in *Proc. of the Intl. Test Conf.*, 2003, pp. 893–901.
4. P. Shivakumar, M. Kistler, S. W. Keckler, D. Burger, and L. Alvisi, "Modeling the effect of technology trends on the soft error rate of combinational logic," in *Proc. of the Intl. Conf. on Dependable Systems and Networks*, 2002, pp. 389–398.
5. M. P. Baze and S. P. Buchner, "Attenuation of single event induced pulses in CMOS combinational logic," *IEEE Transactions on Nuclear Science*, vol. 44, pp. 2217–2223, Dec. 1997.
6. S. Mitra, N. Seifert, M. Zhang, and K. Kim, "Robust system design with built-in soft-error resilience," *IEEE Transactions on Computers*, vol. 38, pp. 43–52, Feb. 2005.
7. M. Nicolaidis, "Time redundancy based soft-error tolerance to rescue nanometer technologies," in *Proc. of the VLSI Test Symposium*, April 1999, pp. 86–94.
8. L. Anghel, D. Alexandrescu, and M. Nicolaidis, "Evaluation of a soft error tolerance technique based on time and/or space redundancy," in *Proc of the Symposium on Integrated Circuits and Systems Design*, Manaus, Brazil, 2000, pp. 237–242.
9. P. Dahlgren and P. Liden, "A switch-level algorithm for simulation of transients in combinational logic," in *Proc. of the Intl. Symposium on Fault-Tolerant Computing*, June 1995, pp. 207–216.
10. A. Kasnavi, J. W. Wang, M. Shahram, and J. Zejda, "Analytical modeling of crosstalk noise waveforms using Weibull function," in *Proc. of the Intl. Conf. on Computer-Aided Design*, 2004, pp. 141–146.
11. W. Chen, S. K. Gupta, and M. A. Breuer, "Analytical models for crosstalk excitation and propagation in VLSI circuits," *IEEE Transactions on Computer-Aided Design of Integrated Circuits and Systems*, vol. 21, no. 10, pp. 1117–1131, Oct. 2002.

12. F. Wang, Y. Xie, R. Rajaraman, and B. Vaidyanathan, "Soft error rate analysis for combinational logic using an accurate electrical masking model," in *Proc. of the Intl. Conf. on VLSI Design*, 2007, pp. 165–170.
13. C. Kashyap, C. Amin, N. Menezes, and E. Chiprout, "A nonlinear cell macromodel for digital applications," in *Proc. of the Intl. Conf. on Computer-Aided Design*, 2007, pp. 678–685.
14. C. Forzan and D. Pandini, "A complete methodology for an accurate static noise analysis," in *Proc. of the Great Lakes symposium on VLSI*, 2005, pp. 302–307.
15. Nanoscale integration and modeling (NIMO) group (2007), *ASU Predictive Technology Model [On-line]*, Available: http://www.eas.asu.edu/~ptm.
16. L. Nagel, "Spice: A computer program to simulate computer circuits," in *University of California, Berkeley UCB/ERL Memo M520*, May 1995.
17. K. Mohanram, "Closed-form simulation and robustness models for SEU-tolerant design," in *Proc. of the VLSI Test Symposium*, 2005, pp. 327–333.

Chapter 4
Modeling Dynamic Stability of SRAMs in the Presence of Radiation Particle Strikes

4.1 Introduction

Static random access memories (SRAMs) are an integral part of modern microprocessors and systems-on-chips (SoCs). Typically, SRAMs occupy more than half of the total chip area [1]. Hence from an economic viewpoint, SRAM yield is very important. In the deep submicron era, the IC supply voltage is often scaled to reduce power consumption [2, 3, 4]. At the same time, noise effects in VLSI designs are increasing [5]. Although voltage scaling reduces the dynamic energy consumption of the IC quadratically, it also reduces the noise tolerance of SRAM cells [3]. Thus SRAM stability analysis has become an essential design task required to improve the yield of processors and SoCs.

Traditionally, static stability analysis was performed for memory designs in nanometer scale technologies. The static noise margin (SNM) [1] is one such metric traditionally used for static SRAM stability analysis. SNM is the maximum amplitude of the voltage deviation on an input node that can be tolerated, without causing a change in the memory state. SNM is obtained by injecting static noise of constant amplitude (for an infinite duration). However, transient noise will typically not be present for a long duration. Also it is possible that a noise of a larger amplitude may lead to a temporary disturbance in the SRAM, but not affect the SRAM state at all. Both the amplitude and the duration of a noise event *together* determine whether the SRAM state will flip or not. However, SNM-based stability analysis fails to capture the time-dependent properties of specific noise events. Hence, the use of SNM to analyze the stability of an SRAM cell (during the design phase) unnecessarily reduces design options, leading to overdesign.

To capture the effects of spectral and time-dependent properties of a noise signal on an SRAM cell, dynamic or time-dependent stability analysis needs to be performed. Dynamic noise margin (DNM) is one such metric for time-dependent stability analysis, which results in a more realistic SRAM noise analysis. However, most of the dynamic stability analysis methods proposed so far involve transistor and device level simulations, which are quite complex and time-consuming in nature [6, 7]. Recently, a model for dynamic stability of SRAMs was reported in [4]. However, this model assumes a rectangular noise signal, which is typically not

R. Garg and S. P. Khatri, *Analysis and Design of Resilient VLSI Circuits: Mitigating Soft Errors and Process Variations*, DOI 10.1007/978-1-4419-0931-2_4,

realistic for practical noise sources. This thus results in a large error compared with SPICE simulations. Therefore, there is a need to develop simple and accurate models for the dynamic stability of SRAMs, which capture the time-dependent nature of the radiation-induced noise signal more closely.

As described in Chap. 1, with technology scaling, radiation particle strikes continue to be problematic for SRAMs. Whether a radiation particle strike (or any other transient noise event) results in the state of SRAM cell being flipped or not depends upon both the amount of charge dumped, as well as the time constants associated with the radiation particle strike. In addition to this, the electric and geometric parameters of an SRAM cell also play an important role in determining whether the SRAM state flips. If the amount of charge dumped by a radiation strike is not sufficient, then the strike will only cause a temporary disturbance and will not result in a state flip. Hence, it is important to develop a compact and accurate model for SRAM cell stability in the presence of radiation particle strikes. Such a model would be very useful to an SRAM designer, allowing them to quickly and accurately evaluate the radiation tolerance of their SRAM cell and make it more reliable. The work presented in this chapter develops a model for the dynamic stability of a 6-T SRAM cell in a holding state, in the presence of radiation events. The model proposed in this chapter can predict the effect of radiation particle strikes quite accurately, and the average error in the estimation of the critical charge of the proposed model is just 4.6% when compared with that of SPICE simulation. The extension of this work to evaluate the dynamic stability of SRAM during other modes (read mode or write mode) of the SRAM cells is straightforward.

The rest of the chapter is organized as follows. Section 4.2 briefly discusses some previous work on modeling the static and dynamic noise margin of SRAMs. In Sect. 4.3, the proposed model for the dynamic stability of an SRAM cell in the presence of a radiation event is described. Experimental results are presented in Sect. 4.4, followed by a chapter summary in Sect. 4.5.

4.2 Related Previous Work

The stability analysis of an SRAM cell has been a topic of great interest for more than a couple of decades, because of its importance in obtaining high yields for microprocessor and SoC designs. A rich and well-developed theory exists for static stability analysis of an SRAM cell [8, 9, 1]. In [9], the authors proved the formal equivalence of four different criterion for worst-case static noise margin. In [1], explicit analytical expressions were presented for the static-noise margin (SNM) as a function of device parameters and supply voltage. Several studies have also been performed to evaluate the effects of process variations on SRAM cell stability, using the SNM [10, 11]. Although a lot of work has been done in static stability analysis, not much work has been reported on the dynamic stability analysis of SRAM cells. Most of the previous work on evaluating the effect of radiation particle strikes (or transient noise) on SRAM stability have used either device level or transistor level simulations [6, 7]. Thus, these methods are time consuming and cumbersome to apply. Recently, in [4], an analytical model for SRAM dynamic stability was presented, for a noise signal consisting of a rectangular current pulse. The authors used nonlinear system theory to derive the equation for the minimum duration of the noise current, which results in the flipping of the SRAM cell state, given the amplitude of the noise current. The authors also attempted to apply their approach to perform transient noise analysis in the presence of radiation particle strikes, but the error of their approach is quite large (11%) compared with SPICE. Also they used a single exponential noise current to model radiation strikes. However, the current due to a radiation particle strike is modeled more accurately by a double exponential [6, 7, 12, 13, 14] current pulse (1.1). In contrast, the model presented in this chapter utilizes a double exponential current pulse (1.1) to model a radiation particle strike, and it is able to predict whether a radiation event will result in a state flip in a 6T-SRAM cell with greater accuracy.

4.3 Proposed Model for the Dynamic Stability of SRAMs in the Presence of Radiation Particle Strikes

In this chapter, a model for the dynamic stability of a 6-T SRAM cell in the presence of a radiation event is presented. Since radiation strikes are random events, when such an event occurs, an SRAM cell will most likely be in a holding state. Thus, the proposed model is presented only for the holding state. However, the extension of the approach presented in this chapter to other states of the SRAM is straightforward.

The approach proposed in this chapter to model the dynamic stability of an SRAM is inspired by the nonlinear system theory based formulation presented in [4]. For brevity, a limited description of the theoretical concepts used in the approach of [4] is provided. Figure 4.1 shows the schematic of a 6-T SRAM cell (note that the access transistors are not shown) with the radiation-induced current ($i_{seu}(t)$) being injected into node $n2$. The total capacitance seen at node $n1$ (and $n2$) is modeled by a capacitor of value C, connected between $n1$ (and $n2$) and ground. The state of the SRAM cell shown in Fig. 4.1 is described by a pair of node voltages (V_{n1}, V_{n2}). From the voltage transfer characteristics (VTC) of Inverter 1 and Inverter 2, the equilibrium points of the SRAM cell are (VDD,GND), (GND,VDD), and (VDD/2,VDD/2). Out of these equilibrium points, (VDD,GND) and (GND,VDD) are the stable equilibria whereas (VDD/2,VDD/2) is a metastable equilibrium. The two VTCs (drawn on the same plot) of the inverters in the SRAM cell also form the state space for the SRAM cell system. In the state space of the SRAM cell, the region of attraction for the (VDD,GND) equilibrium point is the region described by $V_{n1} > V_{n2}$. This means that if the node voltages of the SRAM cell satisfy the condition $V_{n1} > V_{n2}$, then under no external input, the state of SRAM will reach the (VDD,GND) equilibrium point. Similarly, $V_{n1} < V_{n2}$ is the region of attraction for the (GND,VDD) state. The metastable equilibrium point represents the condition

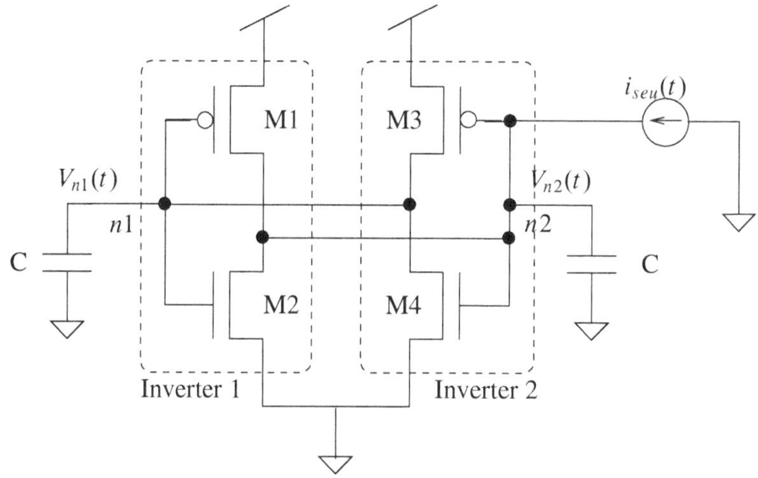

Fig. 4.1 SRAM cell with noise current (access transistors are not shown)

$V_{n1} = V_{n2}$. Therefore, when a SRAM cell is in the metastable state, a small amount of noise at the node $n1$ or $n2$ will drive the state of the SRAM cell to either one of the stable equilibrium points. Thus, a radiation particle strike can flip the SRAM cell state if the radiation-induced current can change the state of the SRAM cell from a stable equilibrium point to the region of attraction of the other stable equilibrium point, or to the metastable point. This criterion is used to evaluate the dynamic stability of the SRAM.

An SRAM cell is a nonlinear system [4] due to the presence of a back-to-back inverter connection. It is very often the case that mathematical tools are unable to analytically solve such nonlinear system equations. Therefore, to ensure that the SRAM dynamic stability model is manageable, a simple linear gate model [15] is used for the inverters. This model assumes that at any given time, either the NMOS or the PMOS device conducts (i.e., the short circuit current of an inverter is negligible). Let the input and output voltages of the inverter be V_{in} and V_{out}. Then the driving current (current flowing through the output node) of an inverter can be written as

$$I_{\text{inv}}(V_{\text{in}}, V_{\text{out}}) = \begin{cases} 0 & \text{cutoff} \\ V_{\text{out}}/R & \text{linear} \\ g_{\text{m}}(V_{\text{in}} - V_{\text{T}}) & \text{saturation} \end{cases}$$

Here, g_{m}, R and V_{T} represent the transconductance, linear-region resistance, and threshold voltage of the transistor of the inverter depending (PMOS or NMOS) which is conducting.

Without loss of generality, the analysis presented in this chapter assumes that initially $V_{\text{in}} = V_{n1} = \text{VDD}$ and $V_{\text{out}} = V_{n2} = \text{GND}$. Therefore, the SRAM cell is in the (VDD, GND) state before the noise current is injected. The same analysis can also be applied when the initial SRAM state is (GND, VDD). Consider the SRAM

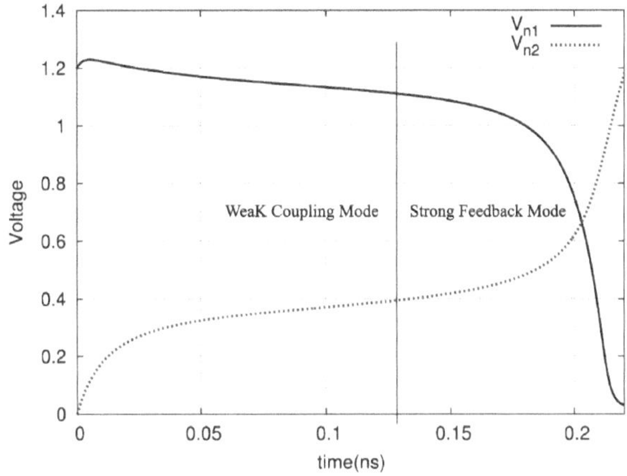

Fig. 4.2 SRAM node voltages for the noise injected at node $n2$

cell of Fig. 4.1. If a sufficiently large noise current is injected into node $n2$, then the SRAM node voltages V_{n1} and V_{n2} change, as shown in Fig. 4.2. Note that Fig. 4.2 is provided for the purpose of explanation only. In practice, the temporal trajectory of V_{n1} and V_{n2} may be different from what is shown in Fig. 4.2, and depends heavily on the value of Q. The goal of the model proposed in this chapter is to test whether a SRAM cell will indeed encounter a state flip, for a given value of Q. Initially, V_{n2} increases. However, V_{n1} remains almost at VDD. Then after the node voltage V_{n2} crosses V_{dsat} (the saturation voltage of NMOS transistor), V_{n1} starts decreasing rapidly. The first phase where V_{n2} is increasing and V_{n1} is constant is referred to as the weak coupling mode (WCM), since the change in V_{n2} does not affect V_{n1}. The second phase, where both V_{n1} and V_{n2} change, is called strong feedback mode (SFM).

Figure 4.3 shows the flowchart of the proposed model, for determining whether a radiation particle strike results in the state of the SRAM cell to flip. The SRAM cell starts in WCM mode when the noise current is injected into node $n2$. If the noise current is sufficiently large, then the SRAM will enter SFM. Otherwise, the SRAM continues to stay in WCM and therefore the SRAM state does not change. After the SRAM cell enters SFM, if the noise current is large enough, then V_{n1} can become greater than or equal to V_{n2}, resulting in a state flip or an SEU. Otherwise, the SRAM cell does not flip and it returns to its initial state (VDD,GND). The steps of the proposed model are explained in detail in the following subsections.

4.3.1 Weak Coupling Mode Analysis

In weak coupling mode, M2 is in the linear region while M4 is in cutoff. M4 is assumed to remain in cutoff during this mode if the threshold voltage of the NMOS

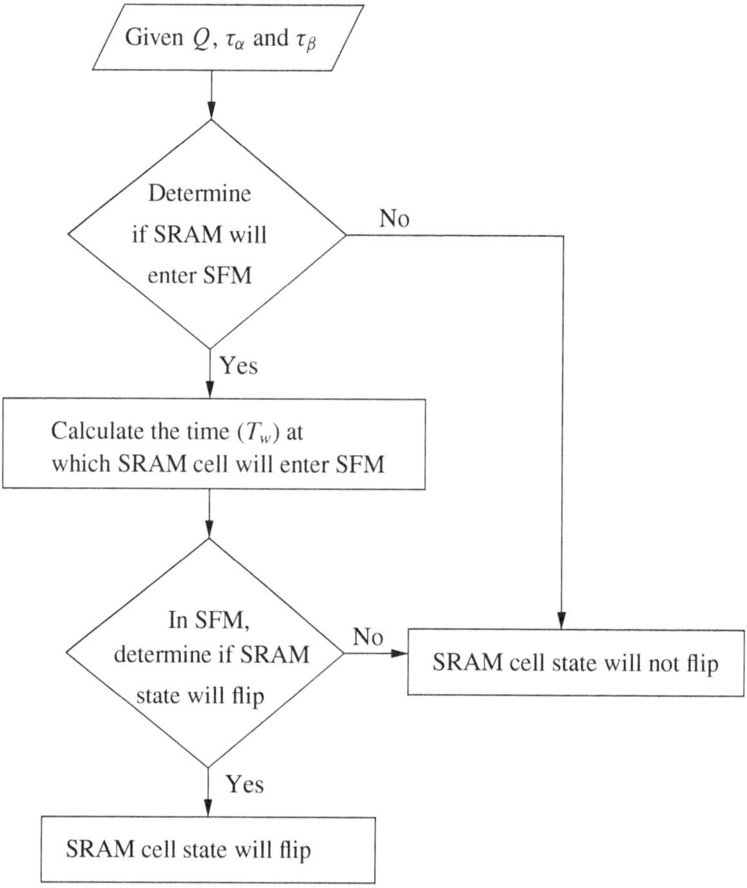

Fig. 4.3 Flowchart of the proposed model for SRAM cell stability

transistor (V_{TN}) does not differ much from V_{dsat}. This is true for deep submicron technologies (due to short channel effects). As mentioned earlier, the node voltage V_{n1} remains almost at VDD in weak coupling mode. The equations governing the temporal behavior of the SRAM cell of Fig. 4.1 are as follows:

$$\mathrm{d}V_{n2}(t)/\mathrm{d}t = -V_{n2}(t)/R_nC + i_{\mathrm{seu}}(t)/C, \tag{4.1}$$

$$V_{n1}(t) = \mathrm{VDD}. \tag{4.2}$$

Here, $i_{\mathrm{seu}}(t)$ represents the radiation-induced current, as described by (1.1). R_n is the linear-region resistance of the NMOS transistor and C is the total capacitance seen at node $n1$ (and $n2$). For a given radiation-induced current pulse, it is required to determine first whether the SRAM cell will enter the strong feedback mode or not. To do this, the minimum value of charge (Q_{wc}) required to take an SRAM cell to

SFM for the given values of time constants (τ_α and τ_β) is computed. To simplify the expression for Q_{wc}, the e^{-t/τ_β} term in the noise current of (1.1) is ignored. Note that e^{-t/τ_β} is ignored only to determine whether the SRAM cell enters SFM or not. Also, ignoring e^{-t/τ_β} results in a pessimistic analysis. Therefore, this assumption results in a lower bounded value of Q_{wc} being computed, and hence does not lead to any error in predicting the SRAM state flip. Integrating (4.1) with initial condition $t = 0$, $V_{n2} = 0$ to obtain:

$$V_{n2}(t) = \frac{Q}{C(\tau_\alpha - \tau_\beta)X}(e^{-t/\tau_\alpha} - e^{-t/R_nC}), \tag{4.3}$$

where

$$X = \frac{1}{R_nC} - \frac{1}{\tau_\alpha}.$$

Now, differentiate (4.3) and equate $dV_{n2}(t)/dt$ to zero, to calculate the time $t_{V_{n2M}}$ at which $V_{n2}(t)$ reaches its maximum value. If $V_{n2}(t_{V_{n2M}}) \geq V_{dsat}$, then the cell enters SFM. Substitute the expression for $t_{V_{n2M}}$ for the value of t and V_{n2} by V_{dsat} in (4.3) to obtain the expression for Q_{wc} as shown below.

$$Q_{wc} = CX(\tau_\alpha - \tau_\beta)\frac{V_{dsat}e^{t_{V_{n2M}}/R_nC}}{e^{t_{V_{n2M}}X} - 1}, \tag{4.4}$$

where

$$t_{V_{n2M}} = \frac{1}{X}\ln\left(\frac{\tau_\alpha}{R_nC}\right).$$

If the charge dumped (Q) by a radiation event is greater than Q_{wc} then the SRAM cell will enter SFM, otherwise it stays in WCM. If the SRAM enters SFM, then the state of SRAM cell *can* flip. To determine whether it indeed flips, it is required to calculate the time (T_w) at which the SRAM cell enters SFM. Again, consider (4.1), integrate it using $i_{seu}(t)$ from (1.1). The resulting equation for $V_{n2}(t)$ is:

$$V_{n2}(t) = \frac{I_n}{C}\left(\frac{e^{-t/\tau_\alpha}}{X} - \frac{e^{-t/\tau_\beta}}{Y} - Ze^{-t/R_nC}\right), \tag{4.5}$$

where

$$Y = \frac{1}{R_nC} - \frac{1}{\tau_\beta}, \quad I_n = \frac{Q}{\tau_\alpha - \tau_\beta} \quad \text{and} \quad Z = \frac{1}{X} - \frac{1}{Y}.$$

To obtain T_w (the time when the SRAM enters SFM), substitute $V_{n2} = V_{dsat}$ and $t = T_w$ in (4.5) and solve it for t. Note that (4.5) is a transcendental equation in t and hence it is not possible to obtain the expression for T_w analytically. Therefore, linearly expand (4.5) in t around the point T_w^{ini} (which is expected to be close to the actual value of T_{w}). To obtain a good expansion point $T_{\mathrm{w}}^{\mathrm{ini}}$, the radiation-induced current is approximated by a rectangular pulse of magnitude $I_{\max}$ (which is the maximum value of $i_{\mathrm{seu}}(t)$) and a pulse width of a $Q/I_{\max}$. Then the value of $T_{\mathrm{w}}^{\mathrm{ini}}$ can be obtained using (4.6) (reported in [4]) for a rectangular noise current pulse for the same SRAM cell as in Fig. 4.1.

$$T_{\mathrm{w}}^{\mathrm{ini}} = -R_n C\, ln[1 - V_{\mathrm{dsat}}/(I_{\max} R_n)]. \tag{4.6}$$

Note that the way in which the radiation-induced current pulse is modeled ensures that $T_{\mathrm{w}}^{\mathrm{ini}}$ is always *smaller* than the actual time (T_{w}) when the SRAM cell enters SFM. This is due to the fact that a rectangular noise current pulse of magnitude $I_{\max}$, depositing a charge Q, has more severe effects on the node voltages than the *actual* radiation-induced current pulse of (1.1) dumping the same amount of charge. It is always better to be conservative so that the SRAM cell state flip is always detected. This ensures that an optimistic SRAM cell design is avoided. Also note from (4.6) that another condition which must be satisfied for an SRAM cell to enter SFM is $I_{\max} R_n > V_{\mathrm{dsat}}$. This condition is checked after the condition imposed by Q_{wc} is satisfied.

To obtain an expression for T_{w}, first linearly expand (4.5) in t around the point $T_{\mathrm{w}}^{\mathrm{ini}}$ (which is obtained from (4.6)) and then solve for t ($= T_{\mathrm{w}}$). The resulting expression for T_{w} is as given below.

$$T_{\mathrm{w}} = T_{\mathrm{w}}^{\mathrm{ini}} + \frac{V_{\mathrm{dsat}} - \frac{I_n}{C}\left(\frac{\mathrm{e}^{-T_{\mathrm{w}}^{\mathrm{ini}}/\tau_\alpha}}{X} - \frac{\mathrm{e}^{-T_{\mathrm{w}}^{\mathrm{ini}}/\tau_\beta}}{Y} - Z\mathrm{e}^{-T_{\mathrm{w}}^{\mathrm{ini}}/R_{\mathrm{nC}}}\right)}{\frac{I_n}{C}\left(-\frac{\mathrm{e}^{-T_{\mathrm{w}}^{\mathrm{ini}}/\tau_\alpha}}{\tau_\alpha X} + \frac{\mathrm{e}^{-T_{\mathrm{w}}^{\mathrm{ini}}/\tau_\beta}}{\tau_\beta Y} + \frac{Z}{R_{\mathrm{nC}}}\mathrm{e}^{-T_{\mathrm{w}}^{\mathrm{ini}}/R_{\mathrm{nC}}}\right)}. \tag{4.7}$$

4.3.2 *Strong Feedback Mode Analysis*

When the SRAM cell of Fig. 4.1 enters strong feedback mode, the transistors M2 and M4 are in the saturation region. In this mode, the node voltage V_{n2} increases (due to the noise current injected at node $n2$) which decreases the value of V_{n1}. The decrease in V_{n1} further helps in increasing the value of V_{n2}. The node voltage V_{n1} depends upon V_{n2} and vice-versa and hence, the equations governing the time-domain behavior of the SRAM cell in the SFM are cross-coupled and non-linear in nature. These equations are given below.

$$\mathrm{d}V_{n1}(t)/\mathrm{d}t = -g_{\mathrm{mn}} V_{n2}(t)/C + g_{\mathrm{mn}} V_{\mathrm{TN}}/C, \tag{4.8}$$

$$\mathrm{d}V_{n2}(t)/\mathrm{d}t = -g_{\mathrm{mn}} V_{n1}(t)/C + g_{\mathrm{mn}} V_{\mathrm{TN}}/C + \frac{Q}{C(\tau_\alpha - \tau_\beta)}(\mathrm{e}^{-t/\tau_\alpha} - \mathrm{e}^{-t/\tau_\beta}). \tag{4.9}$$

Subtracting (4.9) from (4.8) and using transformation $u(t) = V_{n1}(t) - V_{n2}(t)$ gives.

$$\mathrm{d}u(t)/\mathrm{d}t = g_{\mathrm{mn}} u/C - \frac{Q}{\tau_\alpha - \tau_\beta}(\mathrm{e}^{-t/\tau_\alpha} - \mathrm{e}^{-t/\tau_\beta}). \tag{4.10}$$

As mentioned earlier, for the SRAM cell to flip, the noise current should change the state of the SRAM cell from the stable equilibrium point (VDD, GND) to the metastable equilibrium point ($V_{n2} = V_{n1}$), or change the SRAM state to the region of attraction of the other equilibrium point ($V_{n2} > V_{n1}$). Therefore, if the SRAM cell flips, $u(t) = V_{n1}(t) - V_{n2}(t)$ should become equal to or less than 0. Now, integrate (4.10) with the initial condition $t = T_{\mathrm{w}}$ and $u(T_{\mathrm{w}}) = VDD - V_{\mathrm{dsat}}$ and then find the limit of $u(t)$ as $t \to \infty$. This is done because $u(t)$ may become equal to 0 (i.e., $V_{n1} = V_{n2}$) after the entire charge (or most of the charge) has been deposited on node $n2$. Also, the feedback from node $n1$ may also increase the node voltage V_{n2} after a large amount of time. Therefore, the condition which must be satisfied for a radiation event to flip the SRAM state is as given below.

$$Q \geq C(\tau_{\alpha} - \tau_{\beta})\mathrm{e}^{-g_{\mathrm{mn}}T_{\mathrm{w}}/C} \frac{VDD - V_{\mathrm{dsat}}}{\mathrm{e}^{-T_{\mathrm{w}}X'}/X' - \mathrm{e}^{-T_{\mathrm{w}}Y'}/Y'}, \tag{4.11}$$

where

$$X' = \frac{g_{\mathrm{mn}}}{C} + \frac{1}{\tau_{\alpha}}, Y' = \frac{g_{\mathrm{mn}}}{C} + \frac{1}{\tau_{\beta}}.$$

4.4 Experimental Results

To compare the accuracy of the proposed model for the dynamic stability of the SRAM cell with HSPICE [16], the SRAM cell of Fig. 4.1 was designed using a PTM 90 nm [17] model card with VDD $= 1.2\,V$. The device sizes are W/L $=$ 0.18/0.09 μm for M2 and M4 and W/L $=$ 0.27/0.09 μm for M1 and M3. The total node capacitance of nodes $n1$ and $n2$ is 5.4 fF. The gate model characterization (computation of g_{mn}, R_n and V_{dsat}) was done for different VDD values in HSPICE.

As defined in Sect. 1.1, Q_{cri} is the minimum amount of charge required to be deposited by a radiation particle, to flip the SRAM state. Figure 4.4 compares the critical charge values (Q_{cri}) obtained using HSPICE and the model proposed in this chapter (for $\tau_{\alpha} = 150$ ps, $\tau_{\beta} = 38$ ps and for different values of VDD). To obtain the value of Q_{cri}, initially a small value of Q (i.e., 10 fC) is used for the radiation-induced current of (1.1). Then, any of the two methods (HSPICE or the model proposed in this chapter) is used iteratively, with increasing Q in small increments (0.1 fC), to determine the value (Q_{cri}) at which SRAM cell state flips. Figure 4.4 shows that the model proposed in this chapter is very accurate with an average estimation error of 3.3% compared with HSPICE. Thus, the proposed model is much more accurate than the model of [4], whose error is 11%. The error of the model presented in this chapter is lower than that of [4] because unlike [4], the approach of this chapter does not model the radiation-induced current by a rectangular noise current pulse. Hence, the proposed model can capture the time-dependent nature of the radiation-induced current more closely, which improves the accuracy of the model.

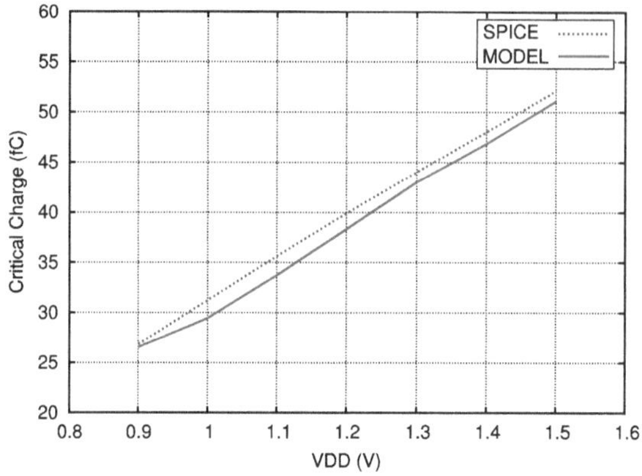

Fig. 4.4 Comparison of critical charge obtained using HSPICE and the proposed model

Table 4.1 Comparison of Model with HSPICE

τ_α	τ_β	VDD	Q_{cri}^{HSPICE} (fC)	Q_{cri}^{MOD} (fC)	% error	Run-time ratio
120	30	1	25.9	24.1	6.95	940
120	30	1.1	29.5	27.2	7.80	1,478
120	30	1.2	33	30.7	6.97	1,811
120	30	1.3	36.5	34.4	5.75	1,830
120	30	1.4	39.8	37.4	6.03	2,058
150	38	1	31.2	29.4	5.77	1,390
150	38	1.1	35.6	33.7	5.34	2,031
150	38	1.2	39.9	38.3	4.01	2,020
150	38	1.3	44	43	2.27	2,488
150	38	1.4	48	46.8	2.50	2,820
150	50	1.1	37.9	36.6	3.43	2,268
150	50	1.2	42.5	41.6	2.12	2,221
150	50	1.3	46.9	46.7	0.43	2,723
AVG					4.57	2,006

Table 4.1 compares the critical charge (Q_{cri}) obtained using the proposed model and HSPICE for various values of τ_α, τ_β and VDD. In Table 4.1, Columns 1 and 2 report the values of τ_α and τ_β under consideration. Column 3 reports the value of VDD. Column 4 reports the critical charge value (Q_{cri}^{HSPICE}) obtained using HSPICE. The critical charge value evaluated by the proposed model (Q_{cri}^{MOD}) is reported in Column 5. Column 6 reports the percentage error in the critical charge value obtained using the model, compared with HSPICE. The ratio of the runtime of HSPICE and the model is reported in Column 7. As reported in Table 4.1, the proposed model is able to obtain Q_{cri} value very accurately (with a small average error of 4.6%). Note that the error of the model reported in [4] was 11%. Also the runtime of the model presented in this chapter is ~2,000× better than the HSPICE runtime. The runtime of HSPICE is in the order of tens of seconds, compared with the runtime of the proposed model, which is the order of 10 ms. Since SRAM design is an

iterative process, it is valuable to use the model proposed in this chapter to evaluate the stability of an SRAM cell due to the significantly lower run-time of this model compared with HSPICE. Note that Fig. 4.4 and Table 4.1 also indicate that the proposed approach is conservative.

4.5 Chapter Summary

SRAMs are extensively used in modern microprocessors and SoCs. Hence SRAM yield is very important from an economic viewpoint. As a result, SRAM stability analysis has become quite important in recent times. SRAM stability analysis based on static noise margin (SNM) often results in pessimistic designs, because SNM cannot capture the transient behavior of the noise. Thus, SNM reduces design options, resulting in a highly conservative design. Therefore, to improve SRAM design, dynamic stability analysis is required. The model developed in this chapter performs dynamic stability analysis of an SRAM cell in the presence of a radiation event. Experimental results demonstrate that the model proposed in this chapter is compact and very accurate, with a low critical charge estimation error of 4.6% compared with HSPICE. The runtime of the proposed model is also significantly lower (2,000× lower) than the HSPICE runtime. Also, the results of the proposed model are always conservative. Thus this model enables SRAM designers to quickly and accurately validate the stability of their SRAMs during the design phase.

The model presented in this chapter considers noise in SRAMs only due to radiation particle strikes. However, there are other types of noise such as power and ground noise, capacitive coupling noise, etc. Therefore, models similar to the one presented in this chapter are required to perform dynamic stability analysis of an SRAM cell in the presence of capacitive coupling noise, and power and ground noise. The approach presented in this chapter could be extended to include the effects of these noise sources as well.

References

1. E. Seevinck, F. J. List, and J. Lohstroh, "Static-noise margin analysis of MOS SRAM cells," *IEEE Journal of Solid-State Circuits*, vol. SC-22, no. 5, pp. 748–754, Oct. 1987.
2. J. Rabaey, *Digital Integrated Circuits: A Design Perspective*, Prentice Hall Electronics and VLSI Series. US Prentice Hall, 1996.
3. K. Takeda, Y. Hagihara, Y. Aimoto, M. Nomura, Y. Nakazawa, T. Ishii, and H. Kobatake, "A read-static-noise-margin-free SRAM cell of low-VDD and high-speed applications," *IEEE Journal of Solid-State Circuits*, vol. 41, no. 1, pp. 113–121, Jan. 2006.
4. B. Zhang, A. Arapostathis, S. Nassif, and M. Orshansky, "Analytical modeling of SRAM dynamic stability," in *Proc. of the Intl. Conf. on Computer-Aided Design*, Nov. 2006, pp. 315–322.

5. S. Rusu, M. Sachdev, C. Svensson, and B. Nauta, "T3: Trends and challenges in VLSI technology scaling towards 100nm," in *Proc. of the Asia South Pacific Design Automation Conf.*, 2002, pp. 16–17.
6. T. May and M. Woods, "Alpha-particle-induced soft errors in dynamic memories," *IEEE Transaction on Electron Devices*, vol. ED-26, pp. 2–9, Jan 1979.
7. J. Pickle and J. Blandford, "CMOS RAM cosmic-ray-induced error rate analysis," *IEEE Transactions on Nuclear Science*, vol. NS-29, pp. 3962–3967, 1981.
8. K. Anami, M. Yoshimoto, H. Shinohara, Y. Hirata, and T. Nakano, "Design consideration of a static memory cell," *IEEE Journal of Solid-State Circuits*, vol. SC-18, no. 4, pp. 414–418, Aug. 1983.
9. J. Lohstroh, E. Seevinck, and J. D. Groot, "Worst-case static noise margin criteria for logic circuits and their mathematical equivalence," *IEEE Journal of Solid-State Circuits*, vol. SC-18, no. 6, pp. 803–807, Dec. 1983.
10. C. Tsai and M. Marek-Sadowska, "Analysis of process variation's effect on SRAM's read stability," in *Proc. of the Intl. Symposium on Quality Electronic Design*, 2006, pp. 603–610.
11. B. H. Calhoun and A. P. Chardrakasan, "Static noise margin variation for sub-threshold SRAM in 65-nm cmos," *IEEE Journal of Solid-State Circuits*, vol. 41, pp. 1673–1679, July 2006.
12. W. Massengill, M. Alles, and S. Kerns, "SEU error rates in advanced digital CMOS," in *Proc. of the European Conf. on Radiation and Its Effects on Components and Systems*, Sep. 1993, pp. 546 – 553.
13. Q. Zhou and K. Mohanram, "Transistor sizing for radiation hardening," in *Proc. of the Intl. Reliability Physics Symposium*, April 2004, pp. 310–315.
14. R. Garg, N. Jayakumar, S. P. Khatri, and G. Choi, "A design approach for radiation-hard digital electronics," in *Proc. of the Design Automation Conf.*, July 2006, pp. 773–778.
15. M. Horowitz, "Timing models for MOS circuits," Ph.D. dissertation, Stanford University, 1984.
16. Synopsys Inc., Mountain View, CA, *HSPICE User's Manual*, 2003.03 edition.
17. Nanoscale integration and modeling (NIMO) group (2007), *ASU Predictive Technology Model [On-line]*, Available: http://www.eas.asu.edu/~ptm

Chapter 5
3D Simulation and Analysis of the Radiation Tolerance of Voltage Scaled Digital Circuits

5.1 Introduction

In addition to the analysis of the effects of radiation particle strikes on combinational circuits and SRAMs, it is also important to study how voltage scaling affects the susceptibility of VLSI circuits to radiation particle strikes. This is relevant since in recent times, power has become a major issue in computing [1]. Low energy solutions are desired for many applications such as systems-on-chip (SoC), microprocessors, wireless communication circuits, etc. Both the dynamic and the leakage components of the power consumption of a CMOS circuit depend upon the supply voltage; both decrease at least quadratically with decreasing supply voltages. Therefore, in recent times, it is common to decrease the supply voltage value in the noncritical parts of VLSI systems, to reduce the power and energy consumption.

Modern VLSI systems extensively employ dynamic voltage scaling (DVS) to meet the variable speed/power requirements that are imposed at different times during their operation [2, 3, 4, 5, 6, 7]. DVS helps in reducing the circuit power consumption especially when high speed circuit operation is not desired. Today, VLSI circuits are also operated in the subthreshold region of operation for a widening class of applications, which demand extreme low power consumption and can tolerate larger circuit delays [8, 9, 10]. Subthreshold circuits operate with a supply voltage less than or equal to the device threshold voltage. Since both DVS and subthreshold circuits are extensively used to reduce power consumption, the susceptibility of such circuits to radiation particle strikes can significantly impact the reliability of VLSI systems based on these techniques. Hence, it is important to analyze the effects of radiation particle strikes on DVS and subthreshold circuits. On the basis of the results of such an analysis, these circuits can be hardened against radiation strikes to improve their reliability.

To understand the effect of voltage scaling on the radiation susceptibility of digital VLSI circuits, in this monograph, 3D simulations of radiation particle strikes (on the output of an inverter implemented using DVS and subthreshold design) were performed. 3D simulation of radiation particle strikes aids in obtaining an accurate estimation of the effect of voltage scaling on the radiation susceptibility of the inverter. A radiation particle strike on an inverter was simulated using Sentaurus-DEVICE [11] for different inverter sizes, inverter loads, supply voltage

R. Garg and S. P. Khatri, *Analysis and Design of Resilient VLSI Circuits: Mitigating Soft Errors and Process Variations*, DOI 10.1007/978-1-4419-0931-2_5,

values (VDD), and the energy of the radiation particles. From these 3D simulations, several nonintuitive observations were made, which are important to consider during radiation hardening of such DVS and subthreshold circuits. On the basis of these observations, several guidelines are proposed for the radiation hardening of such designs, as reported in Sect. 5.4. These guidelines suggest that traditional radiation hardening approaches need to be revisited for DVS and subthreshold designs. A charge collection model for DVS circuits is also proposed, using the results of these 3D simulations. The charge collection model can accurately estimate (with an average error of 6.3%) the charge collected at the output of a gate as a function of the supply voltage, gate size, and particle energy (for medium and high energy particle strikes). The parameters of this charge collection model can be included in transistor model cards in SPICE, to improve the accuracy of SPICE-based simulations of radiation events in DVS circuits.

The rest of the chapter is organized as follows. Section 5.2 discusses some of the previous work in this area. In Sect. 5.3, the 3D simulation setup used for the simulation of a radiation particle strike at the output of inverter is described. In Sect. 5.4, experimental results are presented, and several observations from these results are discussed. The corresponding design guidelines are also presented in this section. Finally, the chapter is summarized in Sect. 5.5.

5.2 Related Previous Work

Although radiation particle strikes in circuits operating at nominal supply voltages have been extensively studied using 3D device simulation tools [12, 13, 14, 15, 16], DVS and subthreshold circuits have not received much attention. In [12], 3-D numerical simulation is used to study the charge collection mechanism in silicon n^+/p diodes. In [17], device level three-dimensional simulation was performed to study the charge collection mechanism and voltage transients from angled ion strikes. The authors of [15] used a 3D device simulation tool to study the effect of radiation-induced transients and estimate the soft error rate (SER) in static random access memory (SRAM) cells. In [18], an experimental study of the effects of heavy ions in commercial SOI PowerPC microprocessors was conducted. Microprocessors implemented using different technology nodes as well as different core voltages were used in the experiment. It was also observed in [18] that the reduction of feature size from 0.18 to 0.13 μm (and core voltage from 1.6 to 1.3 V) had little effect on the soft error rate. The sensitivity of several commercial SRAM devices to radiation, as a function of their supply voltage, was experimentally studied in [19]. An increase in the radiation susceptibility of SRAMs with decreasing supply voltage was observed. The SRAMs used in these experiments were fabricated in older technologies (i.e., the feature sizes were greater than 0.18 μm). The authors of [16] analyzed the dependence of the soft error rate on the critical charge (Q_{cri}) and supply voltage, for a 0.6 μm CMOS process. In both [19] and [16], the study was performed through laboratory experiments, for nominal supply voltage values (the minimum supply voltage value used was 1.5 V in [19] and 2.2 V in [16]). Note that these were not DVS enabled circuits. Hence, the results of [19, 16] cannot be used to predict the susceptibility of DSM VLSI circuits at lower (and subthreshold) voltages. Also, older process technologies were analyzed in [19, 16], and it is expected that circuits implemented with recent deep submicron process technologies can exhibit a very different behavior in response to radiation particle strikes than older processes [14]. In [18, 19, 16], no circuit level radiation hardening guidelines were proposed. In contrast, in the work presented in this chapter, radiation strikes are modeled and analyzed for current technologies, and a set of circuit hardening guidelines are presented based on the findings.

5.3 Simulation Setup

In this work, a radiation particle strike is considered at the NMOS transistor of an inverter (INV) shown in Fig. 5.1. INV is implemented in a 65 nm bulk technology. The input of the INV is at GND and hence, the PMOS transistor is ON and the NMOS transistor is OFF. An industry standard level 3D device simulation (Sentaurus-DEVICE [11]) was used to simulate the INV of Fig. 5.1 with a radiation particle strike at the drain of the NMOS transistor. Sentaurus-DEVICE is a mixed-level device and circuit simulator. The NMOS transistor of the INV was modeled in the 3D device domain as described in Sect. 5.3.1. The PMOS transistor of INV is modeled using a PTM [20] SPICE model (in the circuit domain). Note that a radiation particle strike was not simulated at the PMOS transistor, since it is expected that a particle at the PMOS transistor would yield similar results as obtained from a particle strike at the NMOS transistor.

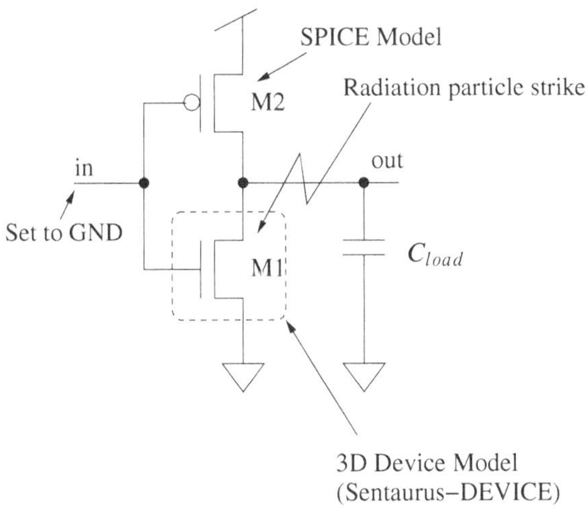

Fig. 5.1 Inverter (INV) under consideration

To analyze the sensitivity of subthreshold circuits, and circuits which employ DVS, to radiation particle strikes, the supply voltage (VDD) of INV was varied in the 3D simulations. The size of INV of Fig. 5.1, as well as the LET of the radiation particle, were varied, to simulate different radiation scenarios. The supply voltage values used were 0.35, 0.5, 0.6, 0.7, 0.8, 0.9, and 1 V. The threshold voltage of the PMOS (NMOS) transistor was $|V_T^P| = 0.365\,V$ ($V_T^N = 0.325\,V$). Hence, 0.35 V was chosen as the supply voltage value for the subthreshold INV. INVs of sizes 2×, 4×, and 15× were simulated. The width of the NMOS (PMOS) transistor in a 2× INV is 0.13 μm (0.52 μm). The INVs were loaded with a load capacitance of value 3× of their input capacitance. The radiation particle LET values used were 2, 10, and 20 MeV-cm^2/mg, which represent low, medium, and high energy strikes, respectively. A 4× INV with LET = 2 MeV-cm^2/mg, and 10 MeV-cm^2/mg, and VDD = $1V$ was also simulated for different load capacitances (0, 1, 3, 5, and 6.3 fF) to study the effect of loading on the radiation susceptibility of the INV.

For each of these simulations, a radiation particle strike was simulated at the center of the drain diffusion of the 3D NMOS transistor. The particle path was along the vertical direction (normal to the surface of the drain diffusion). From simulations, it was found that a vertical strike corresponds to the worst case strike. Hence, in these 3D simulations, the charge collection due to the ALPEN mechanism was not simulated. The total charge collected at the drain node of the INV was due to the drift and diffusion mechanisms, as well as the bipolar effect. The physical models used in the simulations included Shockley-Reed-Hall and Auger recombination, hydrodynamic transport models for electrons, bandgap narrowing dependent intrinsic carrier concentration models, mobility models which included the Philips unified mobility model, as well as high-field saturation and transverse field dependence. The silicon region containing the 3D NMOS device was 10 μm × 10 μm in size.

Note that it is sufficient to model the NMOS transistor in 3D device domain and the PMOS transistor using a SPICE model card to simulate a radiation particle strike. This is because, in an n-well process, the PMOS transistor sits inside an n-well and the n-well terminal is connected to VDD. Therefore, the holes generated by the radiation particle (below the drain of the NMOS transistor in the p-substrate) cannot cross the n-well and p-substrate junction to enter the n-well region. Note that the n-well and p-substrate junction is reverse biased and the n-type diffusion collects only electrons. However, the drain (which is a p-type diffusion) of the PMOS transistor can collect only holes. Thus, the radiation particle strike at the NMOS transistor does not physically affect the PMOS transistor and hence, it is appropriate to use a SPICE model card for the PMOS transistor. Also, when a radiation particle strikes the NMOS transistor, the PMOS transistor is ON since the input of INV is at GND. Because of this, both the source and the drain terminal of the PMOS transistor are at VDD and hence, the drain-bulk junction of the PMOS transistor is not reverse biased. For these reasons, it is common practice to model only the device of a circuit struck by a radiation particle in 3D device domain (the NMOS transistor in this work) [21, 16, 22, 23]. Hence, the approach used in this work is to model the INV is consistent with the previous works.

5.3.1 NMOS Device Modeling and Characterization

The Sentaurus-Structure editor tool [11] was used to construct the 3D NMOS transistor of the INV in Fig. 5.1. The NMOS device was implemented in a 65 nm bulk technology. The 3D 65 nm technology model was developed based on the data available in the literature [1, 24, 25, 26, 27, 22, 23]. On the basis of these references, the value of different parameters used are as follows: the gate length $L = 35$ nm, oxide thickness $T_{ox} = 1.2$ nm, spacer width = 30 nm and the height of the polysilicon gate = 0.12 μm. The threshold voltage, punch through, halo and latchup implants were also modeled in the NMOS device. The details of these implants are as follows. For the threshold (punch through) implant, the peak doping concentration of Boron atoms is $8e^{18}$ cm^{-3} ($7e^{18}$ cm^{-3}) at 2 nm (14 nm) below the SiO_2-channel interface, the doping concentration decreases with a Gaussian profile, and the doping concentration reduces to $1e^{17}$ cm^{-3} ($2e^{17}$ cm^{-3}) at a depth of 14 nm (5 nm) below the peak concentration surface. The peak concentration of Boron atoms for halo implants is $2e^{19}$ cm^{-3}, and these implants are in the channel region at the source-bulk and drain-bulk junctions. Again, the doping concentration reduces with a Gaussian profile. For the latchup implant, the peak doping concentration of Boron atoms is $5e^{18}$ cm^{-3} at 1.25 μm below the SiO_2-channel interface, the doping concentration decreases with a Gaussian profile, and the doping concentration reduces to $1e^{16}$ cm^{-3} at a depth of 0.4 μm. The contact of the p-well was placed at 0.75 μm from the source diffusion of the NMOS transistor. The code that was used to construct the 3D NMOS transistor using the Sentaurus-Structure editor is reported in Appendix A.1. Figure 5.2 shows the NMOS transistor which is simulated in Sentaurus-DEVICE. The cross-section of the NMOS transistor is also shown in the left part of this figure. The 3-D NMOS device constructed in this work was characterized using Sentaurus-DEVICE [11] to obtain the drain current (I_D) as a function of the drain to source voltage (V_{DS}) for different gate to source voltages (V_{GS}). The I–V characteristic of the NMOS transistor with width = 1 μm is shown in Fig. 5.3. Figure 5.3 shows that the NMOS device constructed in this chapter has good MOSFET characteristics. These characteristics were verified to substantially match the 65 nm PTM NMOS device characteristics, using SPICE.

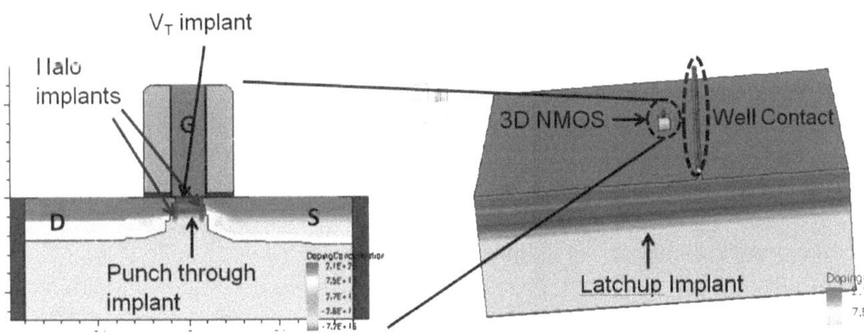

Fig. 5.2 3D NMOS transistor of INV of Fig. 5.1 and its cross-section

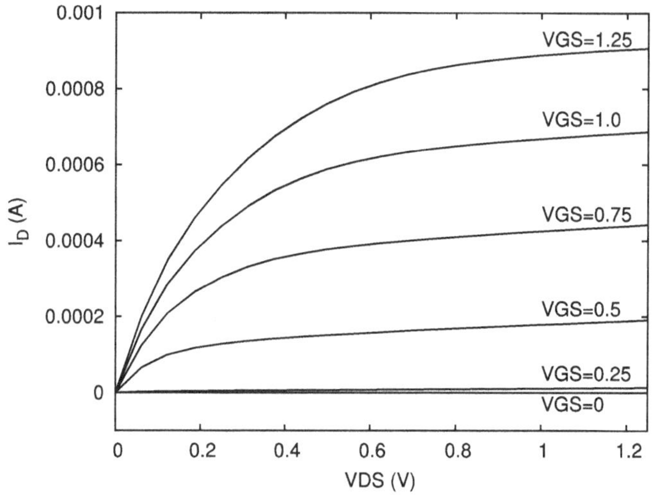

Fig. 5.3 NMOS device: I_D vs. V_{DS} plot for different V_{GS} values

5.4 Experimental Results

Figure 5.4 shows the voltage of the output of the 4× INV of Fig. 5.1 with VDD = $1\,V$, during a radiation particle strike (of three different LET values) at the drain node of the NMOS transistor. Figure 5.7 plots the radiation-induced current through the drain terminal of the NMOS transistor of the 4× INV. Note that for a 65 nm technology, as shown in Fig. 5.4, a radiation particle with an LET value as low as 2 MeV-cm^2/mg is capable of generating a significant voltage glitch (>0.5 VDD). For larger LET values, the voltage at the output of the INV can become negative as shown in Fig. 5.4 (for LET = 10 and 20 MeV-cm^2/mg). Hence, 65 nm devices are very susceptible to radiation particle strikes even with medium energy particles. The radiation-induced voltage transients at the outputs of 2× INV and 15× INV are shown in Figs. 5.5 and 5.6, respectively. From Figs. 5.5, 5.4 and 5.6, it can be observed that upsizing the INV improves its radiation tolerance. From the plots of the radiation-induced NMOS drain current (shown in Fig. 5.7), observe that for low LET values (i.e., 2 MeV-cm^2/mg) the drain current looks like a double exponential current pulse. However, for larger LET values (i.e., 10 and 20 MeV-cm^2/mg), there is plateau in the radiation-induced current. As mentioned in Sect. 1.1, a heavily doped substrate demonstrates charge collection due to both the drift and the diffusion processes. In deep submicron technologies such as 65 nm, the substrate is heavily doped and hence, the funnel collapses very rapidly (within 10–20 ps of the time of the radiation particle strike). As a result, a large amount of charge is left in the substrate (after the funnel collapses), which then gets collected at the drain node of the NMOS transistor through the diffusion process [28]. This results in a significant drain current and hence, the radiation-induced current remains constant for long time. Note that this process is slow, as indicated in [28]. The current plateau was not observed

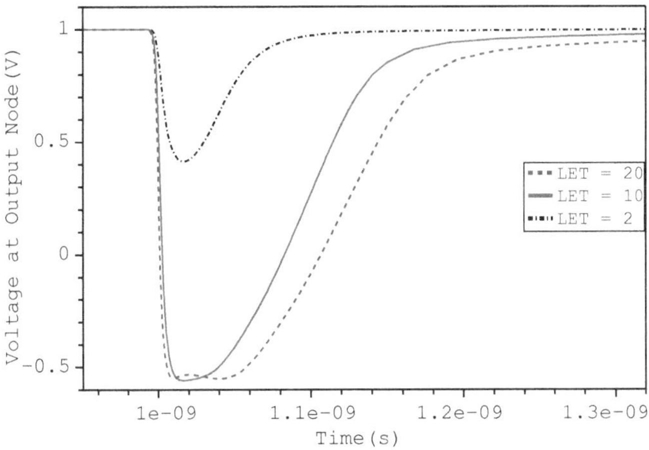

Fig. 5.4 Radiation-induced voltage transient at the output of 4× INV with VDD = 1*V*

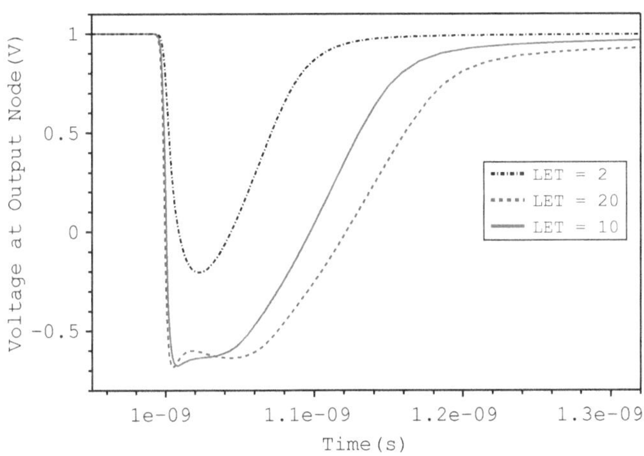

Fig. 5.5 Radiation-induced voltage transient at the output of 2× INV with VDD = 1*V*

for LET = 2 MeV-cm^2/mg since the radiation particle deposits a small amount of charge (20 fC/μm) in the substrate and most of this charge gets collected during the funnel assisted drift collection phase. After this process, very little charge remains in the substrate, which does not result in a significant drain current. The observations made above for the 4× INV from Fig. 5.7 can also be made for 2× INV and 15× INV from Figs. 5.8 and 5.9, respectively. Observe from Fig. 5.9 that the radiation-induced NMOS drain current (for a 15× INV) for LET = 10 MeVcm^2/mg also appears to follow the double exponential model. This is due to the fact that the 15× INV has a strong enough current driving capability so the output voltage remains positive during the radiation-induced transient. Hence, the drain-bulk junction of the NMOS transistor remains reverse biased throughout the radiation-induced transient. As a result, a large portion of the total charge collection occurs by the drift process.

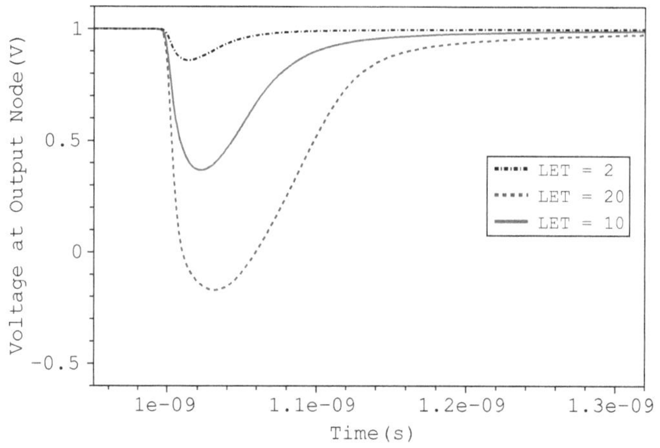

Fig. 5.6 Radiation-induced voltage transient at the output of 15× INV with VDD = $1V$

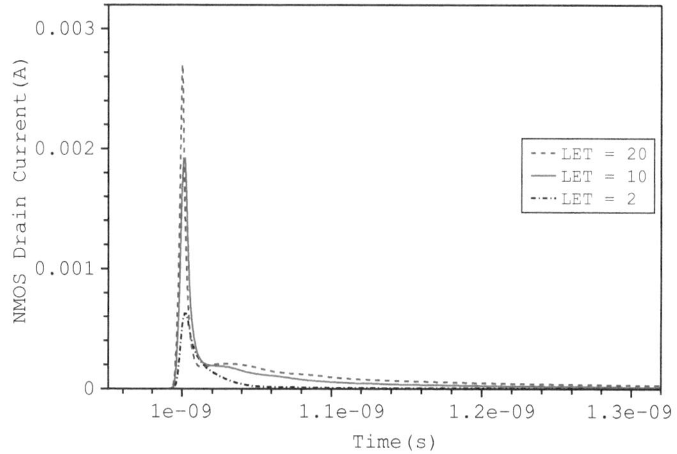

Fig. 5.7 Radiation-induced drain current of the NMOS transistor of 4× INV with VDD = $1V$

The charge collected at the output of INV as a function of the supply voltage during a radiation particle strike is plotted in Fig. 5.10, for different INV sizes and for different linear energy transfer (LETs) values. Figure 5.11 plots the area of the radiation-induced voltage glitch (at the output of INV) for these simulations. Note that in these simulations, the INVs were loaded with a load capacitance of value 3× their input capacitance. The charge collected at the output of the INV is obtained by integrating the drain current of the NMOS transistor following a particle strike. The area of a voltage glitch is computed by integrating the difference of the supply voltage and the voltage at the output of INV ($VDD - V$(out)) following a radiation particle strike. Thus, for a radiation particle strike occurring at time $t1$ at the drain of M1 (shown in Fig. 5.1), the charge collected at *out* is $Q = \int_{t=t1}^{\infty} I_{\mathrm{d}}^{M1}\mathrm{d}t$ and the area of the voltage glitch is $\int_{t=t1}^{\infty}(\mathrm{VDD} - V(\mathrm{out}))\mathrm{d}t$. Note that the area of the

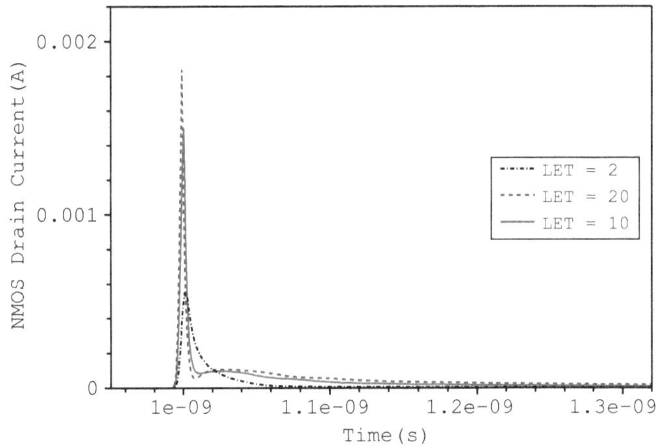

Fig. 5.8 Radiation-induced drain current of the NMOS transistor of 2× INV with VDD = 1V

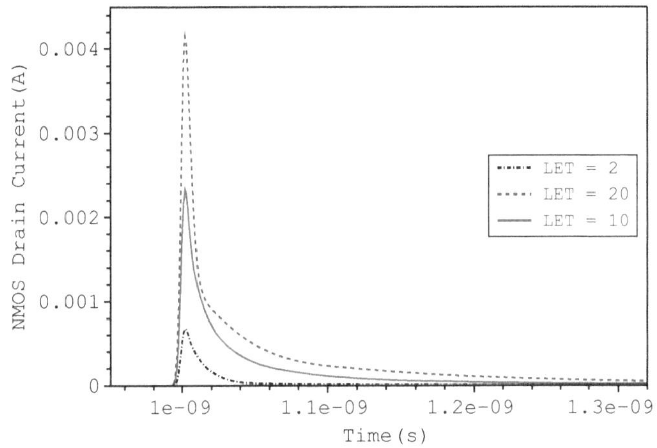

Fig. 5.9 Radiation-induced drain current of the NMOS transistor of 15× INV with VDD = 1V

radiation-induced voltage glitch is a good measure of the susceptibility of an INV (or any gate) to radiation particle strikes, because it incorporates both the magnitude as well as the duration of the voltage glitch. Thus, it can be used for comparison of the susceptibility of INVs across different supply voltage values. From Figs. 5.10 and 5.11 several interesting observations were made. These observations, along with their explanation are as follows.

1. Small devices collect less of the charge deposited by a radiation particle, compared with larger devices. This phenomenon occurs mainly due to two reasons: (1) in a small device, the drain node voltage falls more quickly compared with a large device. Therefore, the strong electric field in the drain-bulk junction of the NMOS exists for shorter duration in the small device than in the large device. Thus, less charge is collected initially during the funnel assisted drift collection

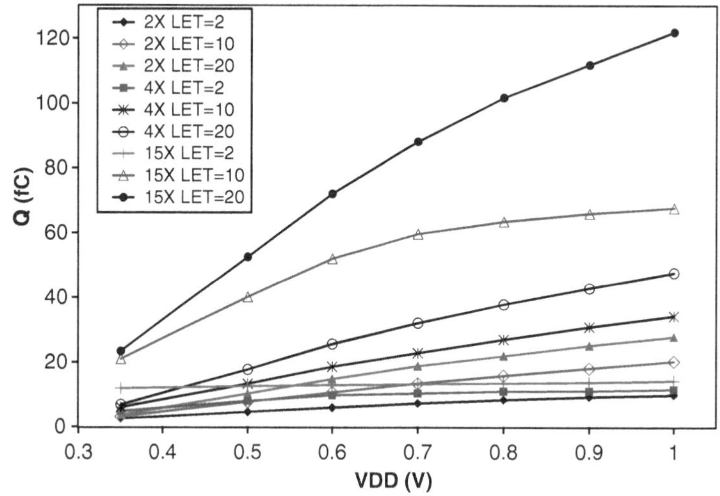

Fig. 5.10 Charge collected at the output of INV for different values

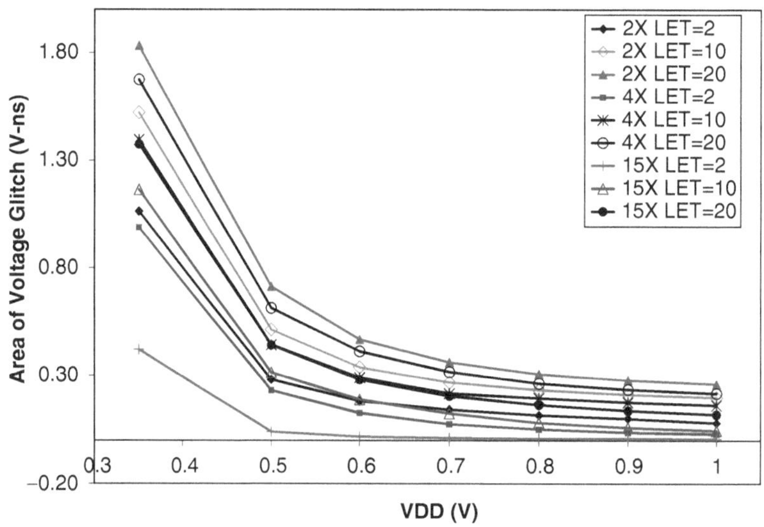

Fig. 5.11 Area of voltage glitch vs. VDD

phase, for a small device. (2) The drain area is smaller in a small device compared with a large device. As a result, less charge is collected through the diffusion process in the small device.

2. For low energy radiation particle strikes, wide devices collect almost the same amount of charge across different supply voltage values. In other words, the charge collection efficiency of wide devices is high and largely independent of supply voltage. As mentioned earlier, during a low energy radiation particle most

of the deposited charge gets collected within a few picoseconds after the particle strike. Also, in a wide device, the drain voltage of the device takes longer to fall, even for low supply voltages, during a low energy radiation strike. Thus, the electric field is present in the drain-bulk junction for a longer duration and a significant amount of charge gets collected, even at low supply voltages.

3. The amount of charge collected due to a radiation particle strike reduces with decreasing supply voltage. The charge collected due to the funnel-assisted drift process depends on the strength of the electric field in the drain-bulk junction. At lower voltages, the electric field in the drain-bulk junction is weaker than at higher voltages. Also, the drain voltage of the device takes longer to fall for higher supply voltages compared with lower supply voltages. Therefore, in case of high supply voltages, the electric field in the drain-bulk junction is strong and present for a longer duration, because of which a large amount of charge gets collected at the drain node (compared with the case when the supply voltage is low).
4. The effects of radiation particle strikes become severe for supply voltages less than 60% of the nominal value (which is slightly lower than the twice of the threshold voltage of the PMOS transistor). As shown in Fig. 5.11, the area of the voltage glitch increases with decreasing supply voltages. The PMOS transistor of the INV is primarily responsible for recovering the voltage at the output node during a radiation particle strike at the NMOS transistor. As the supply voltage (VDD) is decreased, the PMOS transistor drive strength reduces and the PMOS transistor becomes significantly weaker when the supply voltage is reduced below $2 \cdot V_T^P$. Note that the decrease in the drive strength of the PMOS transistor with decreasing VDD value is much higher for VDD $< 2 \cdot V_T^P$ compared with VDD values greater than $2 \cdot V_T^P$. Hence, when the supply voltage is less than $2 \cdot V_T^P$ then the PMOS transistor takes longer to recover the voltage at the output of the INV.

To study the effect of loading on the radiation susceptibility of a gate, a 4× INV with LET $= 2\,$MeV-cm^2/mg and 10 MeV-cm^2/mg, and VDD $= 1\,V$ was also simulated for different load capacitances (0, 1, 3, 5, and 6.3 fF). The results are reported in Table 5.1. In Table 5.1, Columns 1 and 2 report the LET and the load capacitance

Table 5.1 Q and area of voltage glitch vs. load capacitance (C_{load})

LET (MeV-cm^2/mg)	C_{load} (fF)	Q (fC)	Voltage glitch area (V-ns)
2	0	10.0	0.0434
2	1	10.9	0.0448
2	3	11.1	0.0361
2	5	11.3	0.0314
2	6.3	11.6	0.0303
10	0	26.8	0.1224
10	1	27.7	0.1284
10	3	29.9	0.1409
10	5	32.6	0.1549
10	6.3	34.2	0.1629

values under consideration. Column 3 reports the charge collected (Q) at the output of the INV. The area of the radiation-induced voltage glitch is reported in Column 4. From Table 5.1 the following observations were made:

1. For small devices with medium or high energy radiation particle strikes, the pulse width of the voltage glitch *increases* with an increasing load capacitance (C_{load}) of the gate. Because of a radiation particle strike of medium (or high) energy, the voltage at the output of the INV of smaller sizes (such as 4× or smaller) becomes negative very rapidly. After this the PMOS transistor of the INV starts recovering the voltage at the output. If the INV is driving a higher load capacitance (C_{load}), then the PMOS will take a longer time to restore the output voltage. Thus, the width of the voltage glitch increases with the increasing load capacitance (C_{load}), contrary to the popular belief.
2. However, for low energy radiation particle strikes, an increase in the load capacitance (C_{load}) of the gate *improves* the radiation tolerance of the INV. The magnitude of the voltage glitch is reduced with increasing load capacitance (which is due to increasing fanout). This effect is more visible for low energy radiation particle strikes. For high energy strikes, the difference in the magnitude of the voltage glitch for two different loads is very small. As the voltage glitch magnitude is lower for low energy strikes, the PMOS transistor of the INV has to recover a lower voltage swing at the output node. Thus, the width of the voltage glitch reduces with the increasing load capacitance. Hence, the INV becomes more tolerant to low energy radiation strikes with the increasing load capacitance.
3. The charge collected increases with the increasing load capacitance. This is again due to fact that the voltage of the drain node of the NMOS transistor falls slowly for large load capacitances. Thus, the electric field is present in the drain-bulk junction of the NMOS for a longer duration and hence, more charge gets collected.

The observations made above from Figs. 5.10 and 5.11, and Table 5.1 are important to consider during radiation hardening of DVS and subthreshold circuits. On the basis of the observations made, several design guidelines are presented for hardening DVS and subthreshold circuits. These guidelines suggest that the traditional radiation hardening approaches need to be revisited.

1. If a gate is upsized to increase its radiation tolerance, then a higher value of charge collected (due to a radiation particle strike) should be used. This is extremely important for low voltage operation, since lowered voltage circuits are more likely to have large voltage glitch areas.
2. For environments with low energy radiation particles, it is safe to assume that the charge collected remains constant across different supply voltages for wide devices. The collected charge also remains roughly constant across different gate sizes for high or nominal voltage operation.
3. DVS designs should scale down the supply voltage of a circuit to $2 \cdot V_T$ (V_T is the maximum of V_{TP} and V_{TN}). Below this value, radiation susceptibility increases rapidly as shown in Fig. 5.11. Also, a circuit with DVS should be hardened at

the *lowest* operating voltage, with the charge collected at that voltage. This will ensure radiation tolerance at higher supply voltages. Subthreshold circuits and circuits with a supply voltage $<2 \cdot V_T$ require aggressive protection against radiation strikes.

4. The fanout load capacitance (C_{load}) of gates should be kept low in circuits operating in high energy radiation particle environments. This is contrary to conventional wisdom. For low energy radiation environments, the fanout factor (load capacitance) of the gates should be increased to improve their radiation tolerance.

Observe from Fig. 5.10 that the charge collected (Q) at the output of a gate has a strong dependence on the size of the gate, the supply voltage (VDD) and the radiation particle energy (LET). Therefore, simulating radiation particle strikes in DVS circuits in SPICE using a worst case collected charge (maximum possible charge collection) may lead to very pessimistic designs. To improve the accuracy of SPICE simulations of radiation particle strikes in DVS circuits, a model for the charge collected (Q^M) at the output node of a gate is also proposed, and five parameters from this model can be appended in to the SPICE model cards for MOSFETs (e.g., the PTM model cards for 65 nm [20]). Since Q directly depends on the size of a gate – W (expressed in μm), VDD and LET (expressed in MeV-cm^2/mg), the model proposed in this work is $Q^M = \min(K_{MAX} \cdot \text{LET}, K_Q W^{\beta_1} \text{VDD}^{\beta_2} \text{LET}^{\beta_3})$. Here, K_{MAX}, K_Q, β_1, β_2 and β_3 are obtained by characterizing a process technology through 3D simulations of radiation particle strikes. In the expression for Q^M (in fC) $K_{MAX} \cdot$ LET represents the maximum amount of charge that can be collected due to a radiation particle strike. The value of K_{MAX} is obtained from 3D simulations of radiation particle strikes at the drain of a very wide NMOS transistor for different *LET* values. Note that the drain terminal of this NMOS transistor was connected to VDD (nominal value) and the source, gate, and bulk terminals were connected to GND to maximize the charge collection. From 3D simulations, K_{MAX} was found to be 8. Note that since $K_{MAX} = 8$, the amount charge that can be collected (Q) in the worst case is 80% of the charge deposited (Q_D) by a radiation particle strike in the charge collection volume. Therefore, the traditional approach, in which 100% of the charge deposited is assumed to be collected in the worst case, is pessimistic. The values of K_Q, β_1, β_2, and β_3 (parameters in the second term in the expression for Q^M) were estimated by fitting the model Q^M with Q obtained through 3D simulations (shown in Fig. 5.10) for 2×, 4×, and 15× INV, and for VDD = 0.6 to 1.0 V (in steps of 0.1 V) and LET = 10 and 20 MeV-cm^2/mg. The values obtained are $K_Q = 16.54$, $\beta_1 = 0.704$, $\beta_2 = 0.9$, and $\beta_3 = 0.664$. Note that the curve fit was performed for medium and high energy particles, since hardening of DVS circuits needs to be performed against radiation particles of such energies, to meaningfully improve their radiation tolerance. As mentioned earlier, for low energy particle strikes, it is safe to assume that the charge collected remains constant across different supply voltages in wide transistors. Therefore, the Q^M model proposed in this chapter is applicable for medium and high energy particle strikes. Also, as proposed earlier, DVS designs should scale the supply voltages of a circuit up to 60% ($2 \cdot V_T$) of the nominal value, and therefore the Q^M model was obtained for VDD = 0.6–1.0 V. To evaluate the accuracy of the proposed model, the

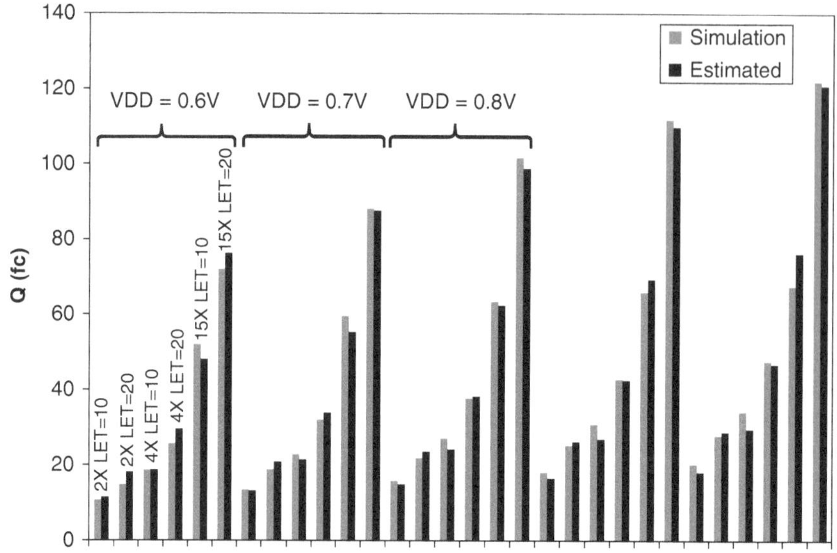

Fig. 5.12 Comparison of charge collected (Q) obtained from the proposed model vs. 3D simulations

amount of charge collected at the output of the INV (shown in Fig. 5.1) predicted by the proposed model (dark bar) and from 3D simulations (light bar) were plotted in Fig. 5.12. Figure 5.12 shows the charge collected with VDD = 0.6–1.0 V in steps of 0.1 V as the outermost variable. For each voltage value, the charge collected is reported for 2×, 4×, and 15× INVs with LET = 10 and 20 MeV-cm^2/mg, as the legend indicates. Figure 5.12 shows that the proposed model is very accurate, with an average error of 6.3%. Thus, the proposed model for the charge collected at the output node of a gate can improve the accuracy of the SPICE level simulation of radiation particle strikes in DVS circuits. For subthreshold circuits, it is difficult to find an accurate model since the charge collection efficiency is very low and hence, 3D simulations should be performed to obtain the value of the charge collected (Q) at the output node of a gate for different parameter values (W and LET).

5.5 Chapter Summary

Radiation particle strikes are becoming increasingly important problems for both combinational and sequential circuits. At the same time, power has become a major issue in computing. In recent times, it is common to decrease the supply voltage value in the noncritical parts of VLSI systems, to reduce the power and energy consumption. Reduced supply voltages further aggravate reliability issues due to radiation. With increasing demand for reliable systems, it is necessary to design radiation tolerant circuits efficiently. In this chapter, the radiation particle strikes in

DVS and subthreshold circuits were studied. 3D simulations for radiation particle strikes in an inverter were performed, using Sentaurus-DEVICE. The sensitivity of DVS and subthreshold circuits to radiation particle strikes were studied by varying the inverter size, the inverter load, the supply voltage (VDD), and the energy of the radiation particle. This was done using 3D simulations. From these 3D simulations, several nonintuitive observations were made, which are important to consider during the radiation hardening of DVS and subthreshold circuits. On the basis of these observations, several guidelines were also proposed for radiation hardening of DVS and subthreshold circuit designs. A model for the charge collected at the output node of a gate was also proposed, which can improve the accuracy of SPICE simulations radiation events.

In the next two chapters, two approaches are presented for hardening a design against radiation particle strikes.

References

1. ITRS, "The international technology roadmap for semiconductors edition," Available: http://public.itrs.net/, 2003
2. R. Gonzalez, B. M. Gordon, and M. A. Horowitz, "Supply and threshold voltage scaling for low power CMOS," *IEEE Journal of Solid-State Circuits*, vol. 32, no. 8, pp. 1210–1216, Aug 1997.
3. T. Burd, T. Pering, A. Stratakos, and T. Brodersen, "A dynamic voltage scaled microprocessor system," in *Proc. of the Intl. Solid State Circuits Conf.*, 2000, pp. 294–295, 466.
4. K. Flautner, S. Reinhardt, and T. Mudge, "Automatic performance setting for dynamic voltage scaling," *Wireless Networks*, vol. 8, no. 5, pp. 507–520, 2002.
5. W. Kim, D. Shin, H. S. Yun, J. Kim, and S. L. Min, "Performance comparison of dynamic voltage scaling algorithms for hard real-time systems," in *Proc. of the Symposium on Real-Time and Embedded Technology and Applications*, 2002, pp. 219–228.
6. B. Zhai, D. Blaauw, D. Sylvester, and K. Flautner, "Theoretical and practical limits of dynamic voltage scaling," in *Proc. of the Design Automation Conf.*, 2004, pp. 868–873.
7. C. Duan and S. P. Khatri, "Computing during supply voltage switching in DVS enabled real-time processors," in *Proc. of the Intl. Symposium on Circuits and Systems*, May 2006, pp. 5115–5118.
8. S. Choi, B. Kim, J. Park, C. Kang, and D. Eom, "An implementation of wireless sensor network," *IEEE Transactions on Consumer Electronics*, vol. 50, no. 1, pp. 236–244, Feb 2004.
9. A. Abidi, G. Pottie, and W. Kaiser, "Power-conscious design of wireless circuits and systems," *Proceedings of the IEEE*, vol. 88, no. 10, pp. 1528–1545, Oct 2000.
10. N. Jayakumar, R. Garg, B. Gamache, and S. P. Khatri, "A PLA based asynchronous micropipelining approach for subthreshold circuit design," in *Proc. of the Design Automation Conf.*, July 2006, pp. 419–424.
11. Synopsys Inc., Mountain View, CA, *Sentaurus user's manuals*, 2007.12 edition.
12. P. E. Dodd, F. W. Sexton, and P. S. Winokur, "Three-dimensional simulation of charge collection and multiple-bit upset in Si devices," *IEEE Transactions on Nuclear Science*, vol. 41, pp. 2005–2017, 1994.
13. P. E. Dodd, "Device simulation of charge collection and single-event upset," *IEEE Transactions on Nuclear Science*, vol. 43, no. 2, pp. 561–575, Apr. 1996.
14. P. E. Dodd and L. W. Massengill, "Basic mechanisms and modeling of single-event upset in digital microelectronics," *IEEE Transactions on Nuclear Science*, vol. 50, no. 3, pp. 583– 602, 2003.

15. J. M. Palau, M. C. Calvet, P. E. Dodd, F. W. Sexton, and P. Roche, "Contribution of device simulation to SER understanding," in *Proc. of the Intl. Reliability Physics Symposium*, March 2003, pp. 71–75.
16. P. Hazucha, C. Svensson, and S. A. Wender, "Cosmic-ray soft error rate characterization of a standard 0.6-μm CMOS process," *IEEE Journal of Solid-State Circuits*, vol. 35, no. 10, pp. 1422–1429, Oct 2000.
17. P. E. Dodd, M. R. Shaneyfelt, and F. W. Sexton, "Charge collection and SEU from angled ion strikes," *IEEE Transactions on Nuclear Science*, vol. 44, pp. 2256–2265, 1997.
18. F. Irom, F. F. Farmanesh, A. H. Johnston, G. M. Swift, and D. G. Millward, "Single-event upset in commercial silicon-on-insulator PowerPC microprocessors," *IEEE Transactions on Nuclear Science*, vol. 49, no. 6, pp. 3148–3155, Dec 2002.
19. O. Flament, J. Baggio, C. D'hose, G. Gasiot, and J. L. Leray, "14 MeV neutron-induced SEU in SRAM devices," *IEEE Transactions on Nuclear Science*, vol. 51, no. 5, pp. 2908–2911, Oct. 2004.
20. Nanoscale integration and modeling (NIMO) group (2007), *ASU Predictive Technology Model [On-line]*, Available: http://www.eas.asu.edu/∼ptm
21. P. E. Dodd, M. R. Shaneyfelt, J. A. Felix, and J. R. Schwank, "Production and propagation of single-event transients in high-speed digital logic ICs," *IEEE Transactions on Nuclear Science*, vol. 51, no. 6, pp. 3278–3284, Dec. 2004.
22. O. A. Amusan, "Analysis of single event vulnerabilities in a 130 nm CMOS technology," M.S. thesis, Vanderbilt University, 2006.
23. S. DasGupta, "Trends in single event pulse widths and pulse shapes in deep submicron CMOS," M.S. thesis, Vanderbilt University, 2008.
24. P. Bai, C. Auth, S. Balakrishnan, M. Bost, R. Brain, and V. Chikarmane et al., "A 65nm logic technology featuring 35nm gate lengths, enhanced channel strain, 8 Cu interconnect layers, low-k ILD and 0.57μm^2 SRAM cell," *International Electron Devices Meeting*, pp. 657–660, Dec. 2004.
25. S. Chen, H. Huang, H. Lin, L. Shie, and M. Chen, "Using simulation for characterize high performance 65nm node planar N-MOSFETs," in *Proc. of the Intl. Symposium on NANO Science and Technology*, Nov. 2004, pp. 1–2.
26. Y. Taur and E. J. Nowak, "CMOS devices below 0.1 um: How high will performance go?," *International Electron Devices Meeting*, pp. 215–218, Dec. 1997.
27. T. Fukai, Y. Nakahara, M. Terai, S. Koyama, and Y. Morikuni et al., "A 65 nm-node CMOS technology with highly reliable triple gate oxide suitable for power-considered system-on-a-chip," *Proc. of the Symposium on VLSI Technology*, pp. 83–84, June 2003.
28. R. Baumann, "Soft errors in advanced computer systems," *IEEE Design & Test of Computers*, vol. 22, no. 3, pp. 258–266, May-June 2005.

Chapter 6
Clamping Diode-based Radiation Tolerant Circuit Design Approach

6.1 Introduction

In Chap. 1, the need to harden combinational circuit was discussed. Then in Chaps. 2, 3, 4 and 5, analysis approaches were presented to analyze radiation-induced transients in combinational circuits and SRAMs. On the basis of the results of the analysis of the effects of a radiation particle strike on a circuit, selective hardening of the gates in a circuit may be performed, to achieve the desired level of radiation tolerance while satisfying area, delay, and power constraints. For this, efficient circuit level hardening techniques are required.

This monograph proposes two circuit level hardening approaches previously published in [1, 2] to harden combinational circuits against a radiation particle strike. These two approaches are referred to as the *diode clamping-based approach* and *the split-output-based hardening approach*. The diode clamping-based approach (which is presented in this chapter) is suitable for hardening combinational circuits against low energy radiation particle strikes. The split-output-based hardening approach (described in the next chapter) is suitable for high energy radiation particle environments.

The diode clamping-based hardening approach is based on the use of shadow gates, whose task is to protect the primary gate in case it experiences a radiation strike. The gate to be protected is duplicated locally, and a pair of diode-connected transistors (or diodes) is connected between the outputs of the original and shadow gates. These diodes turn on when the voltages of the two gates deviate (during a radiation strike). Experimental results show that at the level of a single gate, the area overhead is quite large. At the circuit level, however, gates are selectively hardened. A methodology is presented to protect specific gates of the circuit based on electrical masking, in a manner that guarantees radiation tolerance for the entire circuit. This circuit level hardening methodology is able to harden circuit with low area and delay overheads. An improved circuit level hardening algorithm is also proposed, to further reduce the delay and area overhead.

The remainder of this chapter is organized as follows: Section 6.2 discusses previous work in the area of designing radiation tolerant VLSI circuits. Section 6.3 describes the diode clamping-based radiation tolerant combinational circuit

R. Garg and S. P. Khatri, *Analysis and Design of Resilient VLSI Circuits: Mitigating Soft Errors and Process Variations*, DOI 10.1007/978-1-4419-0931-2_6,

design approach. Experimental results are presented in Sect. 6.4, while the chapter summary is provided in Sect. 6.5.

6.2 Related Previous Work

There has been a great deal of work on radiation hardened circuit design techniques. Several papers focus on combinational circuits [3, 4, 5, 6, 7, 8], while others have focused on memory design [9, 4, 10, 11, 12, 13, 14, 15]. Since memories are particularly susceptible to radiation events, these efforts were crucial to space and military applications.

Circuit hardening approaches can be classified as device level, circuit level, and system level [16, 8, 7, 17, 1, 18, 19, 2]. Device level approaches require fundamental changes to the fabrication process to improve the radiation immunity of a design [16]. Silicon-on-insulator (SOI) devices are considered to be more tolerant than bulk CMOS devices [20] because of the lower charge collection volume in SOI devices. However, other hardening techniques are still required to achieve a meaningful tolerance of an SOI based design to radiation particle strikes [20].

Circuit level hardening approaches use special circuit design techniques that reduce the vulnerability of a circuit to radiation strikes [8, 7, 18, 17, 21, 22, 23]. In [7], the authors selectively upsize the gates in a digital design to increase the radiation tolerance of the design. A larger gate has higher drive capability, which increases its radiation immunity in comparison to a smaller gate. The authors protect those gates in a circuit which contribute maximally to the soft error failure rate of the circuit. These sensitive gates in a circuit are identified by using a logical masking [7] analysis. The authors of [17, 18, 21, 22, 23] also performed selective gate hardening in a circuit. Heijmen et al. performed selective duplication of sensitive gates in [18] (i.e., connecting two gates in parallel) to reduce soft error rate (SER). The authors reported that SER can be improved by 50% with an area penalty of 30%.

Device and circuit level approaches are typically fault *avoidance* approaches, while system level approaches typically involve the use of fault *detection and tolerance* mechanisms. Triple modular redundancy (TMR) [24] is a classical example of a system level design approach. In [10], the authors provide a built-in current sensor (BICS) to detect radiation events in an SRAM, which can be used to trigger a recomputation.

Although the approaches discussed above increase the circuit reliability to radiation events, the cost (in terms of area, delay, and power) associated with these approaches is high, typically unacceptable for high-volume mainstream applications. Also these approaches provide radiation tolerance to radiation particles with moderate energy levels. In other words, the increase in Q_{cri} achieved by traditional approaches is not very high. In several applications, high energy radiation particle strikes are encountered. Therefore, there is a need for a radiation hardening approach, which can provide radiation tolerance against very large values of Q, with comparable or smaller overheads. At the same time, there is a need for radiation hardening approach, which incur low delay penalties for low to medium energy radiation particle strikes.

In this monograph, two circuit level hardening approaches are proposed to harden combinational circuits against a radiation particle strike. The first approach (described in this chapter) *the diode clamping-based approach* is suitable for low energy radiation particle strikes, in circuits which cannot tolerate a large delay overhead due to radiation hardening. The second approach (*the split-output-based hardening approach*) is presented in the next chapter, and it is suitable for high energy radiation particle environments.

6.3 Proposed Clamping Diode-based Radiation Hardening

A radiation strike at a node in a circuit can result in a voltage glitch at that node. If the magnitude of the voltage glitch is more than the switch-point of gates driven by that node, then the radiation-induced transient may propagate to the primary outputs to the circuit. This may result in an SEU. The clamping diode-based circuit hardening approach ensures that such a radiation-induced voltage glitch (at the node where the radiation particle strike occurs) is clamped before it reaches the switch-point of gates in its fanout.

This section is divided into three subsections. In Sect. 6.3.1, two circuit structures (shown in Figs. 6.1 and 6.2), which were investigated to create a radiation-hardened standard cell, are described. Section 6.3.2 discusses the notion of *critical depth* for any protected library cell. A larger critical depth for any cell indicates that more logic stages are needed for this cell to erase the effects of a radiation-induced voltage glitch. On the basis of the notion of critical depth, Sect. 6.3.3 describes two algorithms proposed in this chapter to selectively protect cells in a standard-cell-based circuit, so as to minimize the delay and area overheads.

6.3.1 Operation of Radiation-induced Voltage Clamping Devices

A clamping diode can be used to suppress a glitch. However, this clamping diode should not prevent (or delay) the switching of the logic during its normal functional

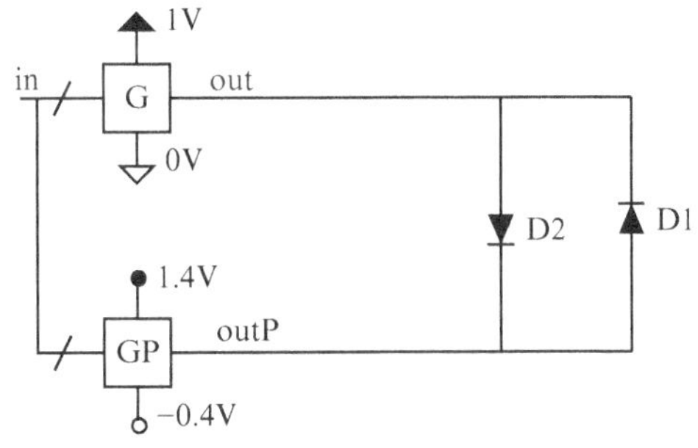

Fig. 6.1 Diode-based radiation-induced voltage glitch clamping circuit

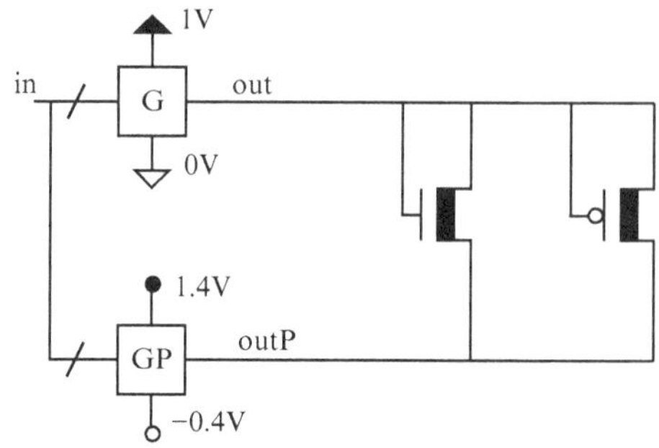

Fig. 6.2 Device-based radiation-induced voltage glitch clamping circuit

operation (when no radiation strike has occurred). Hence, another similarly sized driver (logic gate *GP*) is required in parallel with the gate that is to protect *G*. This is shown in Figs. 6.1 and 6.2. When the outputs of *G* and *GP* deviate significantly (which would occur when one of the gates undergoes a radiation strike), the clamping circuit turns on, thereby protecting the gate *G* from a radiation event. As shown in Figs. 6.1 and 6.2, the supply voltages for the protecting gate (*GP*) are higher (VDD = 1.4 V and VSS = −0.4 V). Hence thicker gate oxides are used for the protecting gate (*GP*) of Figs. 6.1 and 6.2, and the diode connected devices of Fig. 6.2, to avoid reliability problems. Multiple oxide thicknesses have been used in past for a 65 nm process as reported in [25, 26, 27, 28]. Note that it is possible to use a generic CMOS process with a single oxide thickness, but a thicker oxide is required. The thicker oxide will increase short channel effects for the protected gate (*G*), which is powered by VDD and GND. The devices used in the protecting gate have a higher V_T ($V_{TP} = -0.42$ V and $V_{TN} = 0.42$ V) compared with the regular devices used in the protected gate *G* (which have $V_{TN} = 0.22$ V and $V_{TP} = -0.22$ V). This is to minimize the leakage through the protecting gate, which is important since the inputs of *GP* are the same as those of the protected gate. The devices used for clamping also have a higher V_T, to make sure that they are off during regular operation (in the absence of radiation events). In fact the clamping devices are on the verge of conduction (since $V_{TP} = -0.42$ V and $V_{TN} = 0.42$ V). Ideally it would be desired for the protecting gate to have an even higher V_T (to minimize the leakage through this gate), but the proposed circuit hardening approach restricts itself to two V_T values. Note that the bulk terminal of the protecting gate (*GP*) and the diode connected devices of Fig. 6.2 are connected to the protecting gate power supply, i.e., VDD = 1.4 V and VSS = −0.4. This ensures that the bulk terminals of these devices are not forward biased. Also, the dimensions of the devices used in both the hardened and regular version of cells are same. In other words, the sizing of the *G* and *GP* gates in Figs. 6.1 and 6.2 are the same as that of a corresponding unhardened gate.

The clamping diodes used can either be regular p-n junction type diodes (Fig. 6.1) or diode connected devices (Fig. 6.2). Both these options were investigated in the work presented in this chapter. Note that the *Schottky diodes* can also be used as clamping diodes.

6.3.1.1 p-n Junction Diode

Consider the circuit of Fig. 6.1. Assume that a radiation particle strike occurs at the output of protected gate (with its output at logic 0 under steady state), which results in a positive voltage glitch at that node. Note that when the output of the protected gate is at 0 V then the output of the protecting gate is at −0.4 V. When the voltage on the *out* node starts rising and when the voltage across the diode D2 (in Fig. 6.1) reaches the diode turn-on voltage, it begins to clamp the voltage across it. In this way, the glitch due to the radiation event is suppressed.

Now consider the case of a radiation particle strike at the output ($outP$) of the protecting gate which is at logic 0. In this case, the protected node is still protected (remains at logic 0). This is because the protecting node is initially at a much lower voltage (−0.4 V) and as the voltage at the protecting node rises, the diode D2 remains turned-off. Diode D1 turns on only when the voltage at the protecting node rises to a value greater than the diode turn-on voltage (i.e., the voltage glitch magnitude is 0.4 + diode turn-on voltage). However, the radiation particle that can cause such a glitch would have to have a high energy. As mentioned earlier, the proposed clamping diode-based circuit hardening approach is suitable for low energy radiation particle strikes, which cannot result in such a large voltage glitch. Therefore, a low energy radiation particle strike at $outP$ will not affect the voltage at out.

The working of the clamping structure for falling radiation-induced pulses, when the output node is at logic 1, is similar to that discussed earlier.

6.3.1.2 Diode Connected Device

Consider the circuit in Fig. 6.2. Again assume that a radiation event causes a positive voltage glitch at *out* in Fig. 6.2, which was at logic 0 under steady state. At this time, the steady-state output of the protecting gate is at −0.4 V. When the voltage of *out* starts rising, the clamping NMOS device starts to turn on, and conducts more strongly if the voltage of *out* continues to rise, thus clamping the protected node. If the radiation particle strikes the output of the protecting gate, i.e., $outP$, the *out* node remains at logic 0. This is because the protecting node is initially at a much lower voltage (−0.4 V) and as the voltage at the protecting node rises, the clamping NMOS device turns off more. It is only when the voltage of the protecting node rises above 0.4 V that the clamping PMOS device starts turning on. This could cause the voltage of the protected node to rise. As discussed in Sect. 6.3.1.1, a radiation event would need to have a high energy to cause such a glitch.

In a similar manner, the clamping PMOS device helps protect a gate from a falling voltage pulse due to a radiation event.

Both the device-based and diode-based clamping structures were implemented, and had very similar protection characteristics, as shown later in chapter. The layout area penalty of the device-based clamping structure was determined to be lower than that for a diode-based clamping structure. As a consequence, the experiments reported in this chapter are all based on the device-based clamping structure. The performance of device-based and diode-based clamping structures for an inverter are presented in Tables 6.1–6.4.

It was experimentally verified that a radiation strike at the output of the protecting gate does not cause extra soft errors (for the given value of $Q = 24\,\text{fC}$, $\tau_\alpha = 145\,\text{ps}$, and $\tau_\beta = 45\,\text{ps}$). In particular, if there is a radiation particle strike at the output of protecting gate then the Q required to turn on the diode-connected devices and affect the protected node needs to be much larger than 24 fC. Note that the clamping diode-based hardening approach is suitable for low energy radiation particle strikes with Q up to 24 fC. Also, the correct operation of the proposed radiation tolerant gate (shown in Fig. 6.2) was explicitly verified by simulating a radiation particle strike at all nodes of the gates, for every gate in the library (*LIB*) used in this approach to implement radiation tolerant combinational circuits.

6.3.2 Critical Depth for a Gate

Radiation hardened versions for all regular unhardened cells present in the library *LIB* were designed using diode-connected devices. Then the *critical depth* (which is based on the electrical masking) of each radiation hardened cell was computed in the following manner.

Consider a sequence of n copies of the same library cell C, with the output of the ith cell being one of the inputs of the $(i+1)^{\text{th}}$ cell. Let all the other inputs of the $(i+1)^{\text{th}}$ cell be assigned to their noncontrolling values. Assume that a radiation strike occurs at the output of the cell at the first level, with upto $Q = 24\,\text{fC}$, $\tau_\alpha = 145\,\text{ps}$ and $\tau_\beta = 45\,\text{ps}$. Then the critical depth of library cell C, denoted as $\Delta(C)$, is defined as the number of levels of logic that are required for the magnitude of the glitch due to the radiation event to become smaller than $\gamma \times \text{VDD}$, where $\gamma < 1$. Note that $\Delta(C)$ is a function of Q, τ_α, τ_β, the load driven by C and the input ordering of C. The values of $\Delta(C)$, were estimated using SPICE simulations. The worst case critical depth for any library cell C is obtained (by loading it with a single fanout load) in these simulations. Also, for n input gates, the output of each gate was connected to the kth input of the subsequent gates. Then the critical depth was computed as the worst depth among all the n possible input ordering. In this manner, the worst case critical depths was computed for all the cells in the library *LIB*, for the given values of Q, τ_α and τ_β. Note that the definition of critical depth is applicable to static CMOS gates only.

6.3.3 Circuit Level Radiation Hardening

A simplistic approach would be to protect each gate in the design using the standard cell hardening approach proposed in this chapter. However, this would result

Table 6.1 Glitch magnitude of p-n junction clamping diode for rising pulses (output at logic 0)

	Decay time τ_α (ps)								
Q(fC)	105	115	125	135	145	155	165	175	185
21	0.31	0.29	0.28	0.27	0.26	0.24	0.24	0.23	0.22
22	0.33	0.32	0.29	0.28	0.27	0.26	0.25	0.24	0.23
23	0.34	0.32	0.31	0.29	0.28	0.27	0.26	0.25	0.24
24	0.36	0.34	0.32	0.31	0.29	0.28	0.27	0.26	0.25
25	0.38	0.35	0.33	0.32	0.31	0.29	0.28	0.27	0.26
26	0.39	0.37	0.35	0.33	0.31	0.30	0.29	0.28	0.27
27	0.41	0.39	0.36	0.36	0.33	0.31	0.30	0.29	0.28
28	0.43	0.41	0.38	0.36	0.34	0.33	0.31	0.30	0.29
29	0.45	0.42	0.39	0.37	0.35	0.34	0.32	0.31	0.30
30	0.47	0.44	0.41	0.39	0.37	0.35	0.33	0.32	0.31

Table 6.2 Glitch magnitude of p-n junction clamping diode for falling pulses (output at logic 1)

	Decay time τ_α (ps)								
Q(fC)	105	115	125	135	145	155	165	175	185
21	0.31	0.29	0.28	0.26	0.25	0.24	0.24	0.23	0.22
22	0.32	0.32	0.29	0.28	0.26	0.25	0.24	0.24	0.23
23	0.34	0.32	0.30	0.29	0.28	0.26	0.26	0.25	0.24
24	0.35	0.33	0.32	0.30	0.29	0.28	0.27	0.25	0.25
25	0.36	0.34	0.33	0.31	0.30	0.29	0.28	0.26	0.26
26	0.38	0.36	0.34	0.33	0.31	0.30	0.29	0.28	0.26
27	0.40	0.37	0.35	0.34	0.32	0.31	0.30	0.28	0.28
28	0.41	0.39	0.38	0.35	0.34	0.32	0.31	0.29	0.28
29	0.43	0.40	0.38	0.36	0.35	0.33	0.32	0.30	0.29
30	0.45	0.42	0.40	0.38	0.38	0.34	0.33	0.31	0.30

Table 6.3 Glitch magnitude of diode-connected clamping device for rising pulses (output at logic 0)

	Decay time τ_α (ps)								
Q(fC)	105	115	125	135	145	155	165	175	185
21	0.33	0.31	0.29	0.27	0.26	0.25	0.23	0.22	0.21
22	0.36	0.33	0.31	0.29	0.27	0.26	0.25	0.24	0.23
23	0.38	0.35	0.33	0.31	0.29	0.28	0.26	0.25	0.24
24	0.40	0.37	0.34	0.33	0.31	0.29	0.28	0.26	0.25
25	0.42	0.39	0.36	0.34	0.32	0.31	0.29	0.27	0.26
26	0.45	0.41	0.38	0.36	0.34	0.32	0.30	0.29	0.27
27	0.48	0.44	0.41	0.38	0.35	0.34	0.32	0.30	0.29
28	0.50	0.46	0.43	0.40	0.37	0.35	0.33	0.32	0.30
29	0.53	0.49	0.45	0.42	0.39	0.37	0.35	0.33	0.31
30	0.56	0.51	0.47	0.44	0.41	0.39	0.36	0.35	0.33

Table 6.4 Glitch magnitude of diode-connected clamping device for falling pulses (output at logic 1)

	Decay time τ_α (ps)								
Q(fC)	105	115	125	135	145	155	165	175	185
21	0.32	0.30	0.28	0.27	0.26	0.24	0.23	0.22	0.21
22	0.33	0.31	0.30	0.28	0.26	0.25	0.24	0.23	0.22
23	0.35	0.33	0.31	0.29	0.28	0.27	0.25	0.24	0.23
24	0.38	0.35	0.33	0.31	0.29	0.28	0.26	0.25	0.24
25	0.40	0.37	0.34	0.33	0.31	0.29	0.28	0.26	0.25
26	0.41	0.38	0.36	0.34	0.32	0.30	0.29	0.28	0.27
27	0.43	0.41	0.38	0.35	0.34	0.32	0.30	0.29	0.28
28	0.45	0.43	0.40	0.37	0.35	0.33	0.32	0.30	0.29
29	0.48	0.44	0.42	0.39	0.37	0.35	0.34	0.31	0.30
30	0.50	0.46	0.43	0.40	0.38	0.36	0.34	0.33	0.31

in an exorbitant delay and area overhead for the circuit. Instead, a selective hardening approach in presented where the delay and area overhead is minimized, while guaranteeing radiation hardness for the circuit.

Let $\Delta = \max_C(\Delta(C))$ over all the cells in the library *LIB*. Given any circuit, one could protect all gates that are topologically Δ or less levels away from any primary outputs of the circuit. In this case, if there is a radiation strike on any protected cell, it would be eliminated because the cell is protected. If there is a radiation strike on an unprotected cell, it would be eliminated since the effect of the strike needs to traverse Δ or more levels of protected gates before it reaches the output. In either case, the circuit is tolerant to the radiation event.

A variant of the above approach, which is slightly more efficient, is based on *variable depth protection*, and is described in Algorithm 1. It is based on a reverse topological traversal of a circuit η from its primary outputs. Let *deptharray*() be the array of critical depths of all the library cells used in the implementation of the circuit η. The algorithm starts with a requirement to protect gates up to a reverse topological depth $D = \Delta(p)$, where $\Delta(p)$ is the critical depth of the gate at the primary output p. Whenever a gate C with critical depth $\Delta(C)$ is encountered, the algorithm updates the depth to be protected as $D = \min(D - l, \Delta(C))$. Here, l is the topological depth of gate C from the primary output p.

6.3.4 *Alternative Circuit Level Radiation Hardening*

If a large number of gates with high critical depth are present near the primary outputs of a circuit, then it might be necessary to protect a significant portion of the circuit using the variable depth protection approach. This will result in large area and delay overheads. Column 8 of Table 6.5 reports the critical depth of all the gates

Algorithm 1 Variable Depth Radiation Hardening for a Circuit

variable_depth_protect (η, *deptharray*)
for *each* $p \in PO(\eta)$ **do**
  $D = \Delta(p)$
  for *each cell C such that* $p \in fanouts(C)$ **do**
    l = topological depth of C from p
    $D = min(D - l, \Delta(C))$
    if $D > 1$ **then**
      Replace C by $C_{hardened}$
    end if
  end for
end for

in the library *LIB*. Observe from this table that the critical depth of the inv2AA gate is much higher than the rest of the gates in *LIB*. Therefore, if a large number of inv2AA gates are present near the primary outputs of a circuit, then the area and delay overhead of the hardened circuit will be large. Thus, to further reduce the area and delay overhead associated with variable depth protection scheme, an algorithm is presented, which attempts to reduce the number of gates with large critical depth (such as inv2AA) near the primary outputs of a circuit.

The proposed approach to further reduce the area or delay overhead is described in Algorithm 2. Let η be a mapped circuit obtained using library *LIB* with either area or delay as a cost function. Also let η^* be the circuit obtained after using the variable depth protection algorithm on η. Now, partition η^* into two parts, where the first part is the unprotected portion of η^*, represented by ζ, and the second part is the protected portion of η^*, represented by ϕ. Then modify the library *LIB* to obtain another library L^* in which a large area and delay cost are assigned to gates with

Table 6.5 Delay, area, and critical depth of cells

Cell	Reg. delay (ps)	Hard. delay (ps)	Delay % ovh.	Reg. area (μm^2)	Hard. area (μm^2)	Area % ovh.	Depth
inv2AA	24.04	26.24	9.16	1.53	8.15	433.33	4
inv4AA	23.91	22.75	−4.88	2.04	9.60	370.83	1
nand2AA	31.42	33.01	5.06	2.04	9.17	350.00	1
nand3AA	44.92	46.10	2.63	2.55	10.70	320.00	1
nand4AA	62.44	63.34	1.44	3.06	12.23	300.00	1
nor2AA	45.62	48.46	6.24	2.55	10.19	300.00	2
nor3AA	77.15	81.04	5.04	4.59	14.52	216.67	1
nor4AA	92.80	92.74	−0.07	7.13	18.86	164.29	1
and2AA	57.48	58.52	1.81	2.55	10.19	300.00	1
and3AA	76.90	75.67	−1.60	3.06	11.72	283.33	1
and4AA	98.75	99.60	0.86	3.57	12.74	257.14	1
or2AA	71.16	71.00	−0.23	3.57	12.23	242.86	1
or3AA	112.87	113.37	0.44	5.35	15.29	185.71	1
or4AA	125.17	123.51	−1.32	8.15	20.89	156.25	1
AVG			1.76			277.17	

large critical depths (for example inv2AA). Resynthesize ϕ with the new library L^* to obtain ϕ^*, which will contain very few gates of high critical depth because of the high cost associated with them. Then, append ζ to ϕ^* and apply the variable depth protection algorithm on the combined circuit to produce a radiation tolerant circuit η'. The resulting circuit η' as is referred to as the *resynthesized hardened circuit* in the sequel.

Algorithm 2 Alternative circuit level radiation hardening

alternative_circuit_protect (η, L, *deptharray*)
$\eta^* = variable\_depth\_protect(\eta, deptharray)$
Partition η^* *into* (ζ,ϕ)
$L^* = modify(L)$
$\phi^* = re - synthesize(\phi, L^*)$
$\eta^c = append(\zeta, \phi^*)$
$\eta' = variable\_depth\_protect(\eta^c, deptharray)$

6.3.5 *Final Circuit Selection*

Two different radiation tolerant versions η^* and η' of a regular circuit η are obtained using the approaches described in Sects. 6.3.3 and 6.3.4. Now obtain the delay and area associated with both η^* and η'. The final radiation tolerant circuit can be obtained by choosing η^* or η' such that the area or the delay is minimized. This approach is referred to as the *improved circuit protection approach.*

6.4 Experimental Results

The radiation tolerance of both radiation hardened gate structures shown in Figs. 6.1 and 6.2 was simulated in SPICE [29]. A 65 nm BPTM [30] model card was used, with VDD = 1 V and $V_{TN} = |V_{TP}| = 0.22$ V.

On the basis of [7], $\tau_\beta = 45$ ps was used. The value of τ_α and Q was varied, to test the proposed radiation hardened gate design against a variety of radiation conditions.

The performance of both radiation hardened gate designs is summarized in Tables 6.1–6.4. These tables report the protection results (in terms of the magnitude of the radiation-induced voltage glitch) for the (inv2AA) gate, which is the most radiation sensitive gate in the library *LIB*. The first two tables report the simulation results for diode-based clamping, and the latter two describe the results for device-based clamping. For both styles, the glitch magnitude is reported for varying values of τ_α and Q. The first and third tables report values of the glitch magnitudes when the output is at logic 0, while the second and fourth correspond to an output at logic 1.

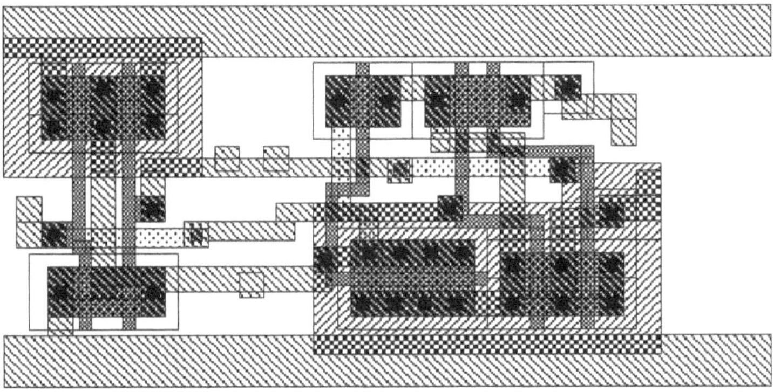

Fig. 6.3 Layout of radiation-tolerant NAND2 gate (uses device based clamping)

On the basis of these tables, it can be observed that the regular p-n junction diode tended to have better protection performance than the diode connected device for the same active area. However, implementing the p-n junction diodes require a larger area on account of the spacing requirements of the required wells, which are at different potentials. The diode connected devices share their well with the devices in the protecting gate, and can be implemented in a more area-efficient manner. Also, the leakage current of the regular p-n junction diode will be higher under delay variations, which can lead to a large voltage drop across the diode. Therefore, the diode-connected devices of Fig. 6.2 were used for hardening gates. Note that the Schottky diodes can be used instead of regular p-n junction diodes. The Schottky diodes can be implemented in smaller area compared with the regular p-n junction diodes. The library *LIB* consists of inv2AA, inv4AA, and2AA, and3AA, and4AA, or2AA, or3AA, or4AA, nand2AA, nand3AA, nand4AA, nor2AA, nor3AA, and nor4AA gates. Layouts were created for both the hardened and regular versions of all the gates in the standard cell library *LIB*. Figure 6.3 describes the layout of the device-based clamping approach, for the nand2AA.

Figure 6.4 describes the voltage waveform at the output of inv2AA, when radiation particle strikes corresponding to $Q = 24\,\text{fC}$, $\tau_\alpha = 145\,\text{ps}$, and $\tau_\beta = 45\,\text{ps}$ were simulated at its output node. The voltage waveform of the unprotected design experiences a large glitch. If it were captured by a memory element, an incorrect value would be sampled. The proposed device clamping-based hardened inv2AA successfully clamps the voltage to a safe level.

Figure 6.5 shows the voltage waveform at the output of a gate, when a current corresponding to $Q = 24\,\text{fC}$, $\tau_\alpha = 145\,\text{ps}$, and $\tau_\beta = 45\,\text{ps}$ is injected into *the protecting* node. The voltage waveform of the output node is well within the noise margins of the gate.

Based on the fact that the device-based protection scheme is used due to its reduced layout area compared with the diode clamping approach, the largest value of Q that the inv2AA cell can tolerate (from Tables 6.3 and 6.4) is 24 fC for $\tau_\alpha = 145\,\text{ps}$. This corresponds to $\gamma = 0.35$ (i.e., the designs can tolerate a glitch magnitude of $0.35 \times \text{VDD}$).

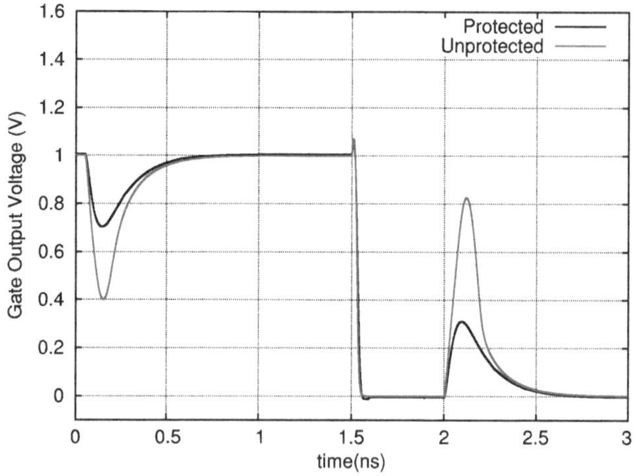

Fig. 6.4 Output voltage waveform during a radiation event on output

Fig. 6.5 Output voltage waveform during a radiation event on protecting node

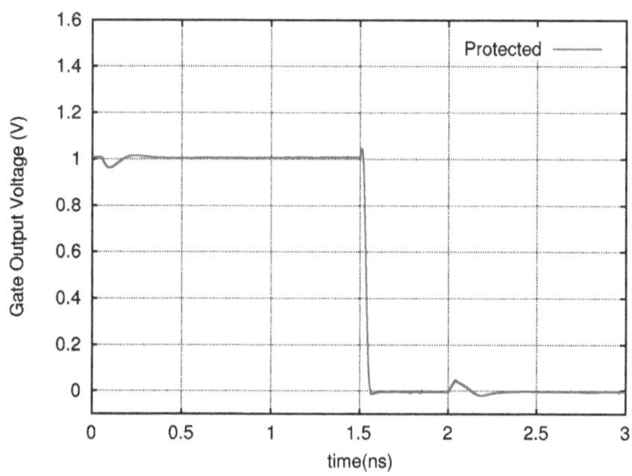

Based on the values of $Q = 24\,\text{fC}$, $\tau_\alpha = 145\,\text{ps}$, and $\tau_\beta = 45\,\text{ps}$, the critical depth $\Delta(C)$ for each gate C in *LIB* was computed. The results of this exercise are presented in Table 6.5, in Column 8. In addition to critical depth, Table 6.5 also reports the worst-case delay (in picoseconds) and the layout area (in μm^2) of each cell in *LIB*. Columns 2 and 3 report the worst case delay of the unprotected and protected versions of the cell. Column 4 reports the percentage overhead in the worst-case delay of the hardened version of each cell compared with its regular version. Note that the worst-case delay of the protected cell is on an average just slightly larger than that of a regular cell. Also note that for some cells (inv4AA, and3AA, etc.), the delay overhead is negative. This is possibly due to the fact that the leakage current of the hardened version of those cell is greater than the regular cell, therefore resulting in faster output transitions. Columns 5 and 6 report the layout area of unprotected and protected versions of cells. The area overhead of hardened version of each cell compared with its regular version is reported in Column 7.

Note that the average area overhead is about 277%, which is quite large. Therefore, the variable depth protection and improved variable depth protection algorithms are used to harden a circuit, so that only a few gates are replaced with their radiation tolerant version. This helps in achieving a reduced area overhead.

Table 6.6 reports the delay overhead of the proposed circuit hardening approaches (η^* and η') for both area and delay mapped designs. The area overhead of the radiation tolerant approaches is reported in Table 6.7. Tables 6.9 and 6.10 report the delay and the area overhead, respectively, of the best radiation tolerant circuit (between η^* and η') using delay or area-based mapping. The circuits were optimized using technology independent optimization in SIS (including redundancy removal), and were then mapped for area and delay using the 65 nm standard cell library *LIB*.

The delay penalty associated with applying the proposed radiation hardening approaches (η^* and η') is presented in Table 6.6. Delays were computed using the *sense* [31] package in SIS [32], which computes the largest sensitizable delay for a mapped circuit. In Table 6.6, Columns 2 and 3 report the delay (in picoseconds) of a regular design and a radiation-hardened area-mapped design (before resynthesis). Column 4 reports the percentage delay overhead for the radiation-hardened design. Column 5 reports the delay of resynthesized radiation-hardened area-mapped design (which are obtained as described in Sect. 6.3.4) and Column 6 reports the percentage delay overhead for this design. Similarly, Columns 7 and 8 report the delay (in picoseconds) of a regular design and a radiation-hardened delay-mapped design (before resynthesis). Column 9 reports the percentage delay overhead for the radiation-hardened design. Column 10 reports the delay of the resynthesized radiation-hardened delay-mapped design, and Column 11 reports the percentage delay overhead for this design. As reported in Table 6.6, the circuit-level delay overhead of the variable depth protection algorithm is as low as 2.92% on average for delay mapped designs, and about 1.6% for area mapped designs (before resynthesis). Note that the radiation-hardened designs are generated by replacing regular gates (which are topologically close to the outputs) by hardened gates. This results in a large increase in the load capacitance of the regular gates that drive these hardened gates. As a consequence, the circuit level delay penalty in Table 6.6 is sometimes larger than the gate-level delay penalty reported in Table 6.5. The circuit-level delay overhead of the resynthesized hardened circuit is 2.63% on average for delay mapped designs, and about 8.11% for area mapped designs, which is higher than the delay associated with hardened circuit before resynthesis. For area mapped circuits, the delay overhead increases (for η') because, for resynthesis of the hardened circuit, first the hardened portion of the circuit obtained from the variable depth protection algorithm is extracted. Then this subcircuit is resynthesized with a high cost assigned to gates with a large critical depth, to minimize their utilization. This increases the utilization of gates with a large input load capacitance and hence, the load on the unprotected circuit increases, resulting in a delay increase. However, for delay mapped designs, the delay overhead reduces due to the increased usage of low overhead (and negative overhead) gates. Also note that in some circuits, the delay overhead of the hardened circuit is negative. This is due to the increased usage of

Table 6.6 Delay overhead of the proposed radiation hardened design approaches

	Area mapping					Delay mapping				
Ckt	Regular	η^*	%Ovh.	η'	%Ovh.	Regular	η^*	%Ovh.	η'	%Ovh.
alu2	1211.680	1165.100	−3.84	1214.939	0.27	1052.595	1066.158	1.29	1073.261	1.96
alu4	1405.975	1435.371	2.09	1533.967	9.10	1319.840	1329.837	0.76	1425.119	7.98
C1355	960.003	990.448	3.17	984.751	2.58	775.568	787.417	1.53	787.417	1.53
C1908	1376.626	1385.880	0.67	1486.142	7.96	1172.012	1184.320	1.05	1215.548	3.71
C3540	1682.691	1728.315	2.71	1772.920	5.36	1560.553	1571.991	0.73	1588.231	1.77
C499	960.003	990.448	3.17	984.751	2.58	775.568	787.417	1.53	787.417	1.53
C880	1606.093	1669.115	3.92	1323.711	−17.58	1544.077	1570.779	1.73	1239.997	−19.69
dalu	1325.516	1363.747	2.88	1415.225	6.77	1221.374	1233.771	1.02	1232.717	0.93
des	2170.999	1721.303	−20.71	2595.902	19.57	2016.371	2053.416	1.84	2272.788	12.72
frg2	910.514	930.828	2.23	991.758	8.92	911.745	957.092	4.97	870.592	−4.51
i2	462.161	477.990	3.42	478.714	3.58	377.435	386.718	2.46	417.151	10.52
i3	172.459	199.782	15.84	233.170	35.20	172.459	199.782	15.84	194.383	12.71
i10	2217.855	2335.109	5.29	2685.245	21.07	2246.170	2318.547	3.22	2315.502	3.09
AVG			1.60		8.11			2.92		2.63

Table 6.7 Area overhead of the proposed radiation hardened design approaches

	Area mapping					Delay mapping				
Ckt	Regular	η^*	%Ovh.	η'	%Ovh.	Regular	η^*	%Ovh.	η'	%Ovh.
alu2	1045.88	1418.28	35.61	1215.22	16.19	1397.26	1569.74	12.34	1569.74	12.34
alu4	1994.52	2470.09	23.84	2279.11	14.27	2470.09	2756.25	11.59	2756.25	11.59
C1355	1592.01	2121.52	33.26	1994.52	25.28	1728.90	2279.11	31.82	2279.11	31.82
C1908	1569.74	1994.52	27.06	1799.46	14.63	1799.46	2225.95	23.70	2279.11	26.66
C3540	3183.22	3916.26	23.03	3573.65	12.27	4022.10	4572.46	13.68	4515.84	12.28
C499	1569.74	2121.52	35.15	1994.52	27.06	1728.90	2279.11	31.82	2279.11	31.82
C880	1045.88	1752.26	67.54	1418.28	35.61	1397.26	1871.43	33.94	1764.00	26.25
dalu	2470.09	2996.47	21.31	2965.89	20.07	3310.85	4057.69	22.56	3573.65	7.94
des	9964.03	16842.85	69.04	13731.15	37.81	12139.63	17800.90	46.63	15490.29	27.60
frg2	1994.52	4201.63	110.66	3916.26	96.35	2611.21	4147.36	58.83	4238.01	62.30
i2	685.39	730.08	6.52	745.29	8.74	872.61	872.61	0.00	872.61	0.00
i3	495.51	670.81	35.38	600.25	21.14	495.51	656.38	32.47	600.25	21.14
i10	6037.29	12016.54	99.04	9304.53	54.12	7705.33	11231.76	45.77	11054.42	43.46
AVG			45.19		29.50			28.09		24.25

the hardened inv4AA gate, which has a negative delay overhead over the regular inv4AA gate.

Both the regular and the radiation hardened circuits were mapped using the library of cells mentioned in the beginning of this section. The resulting designs were placed and routed using SEDSM [33]. Note that the area overhead due to the routing of the additional power supplies has been accounted for. The additional supply lines (VDD $= 1.4$ V and GND $= -0.4$ V) were routed as regular signal lines. This was done because a single radiation particle strike would result in the clamping action at only one gate in an entire circuit and therefore, wider wires are not needed for additional supply lines. The area penalty associated with applying the proposed protection algorithms (η^* and η') is presented in Table 6.7. In Table 6.7, Columns 2 and 3 report the placed-and-routed area (in μm^2) of a regular design and the radiation-hardened area-mapped design (before resynthesis). Column 4 reports the percentage area overhead for the radiation-hardened design. Column 5 reports the placed-and-routed area of the resynthesized hardened area-mapped design, and Column 6 reports the percentage area overhead for this design. Similarly, Columns 7 and 8 report the area (in μm^2) of a regular design and a radiation-hardened delay-mapped design (before resynthesis). Column 9 reports the percentage area overhead for the radiation-hardened design. Column 10 reports the placed-and-routed area of the resynthesized radiation tolerant delay-mapped design, and Column 11 reports the percentage area overhead for this design. Observe from Table 6.7 that the area overheads on average are larger for area-mapped designs, which is reasonable since the designs were mapped with an area-based cost function to start with. The average area penalty was about 45% and 28% for area and delay mapped designs obtained using variable depth protection approach before resynthesis. However, the area overhead was around 29% and 24% for the resynthesized area and delay mapped hardened designs. The area overhead of the resynthesized designs is lower than that of the original designs, since a small number of gates with high critical depth are used in the resynthesized circuit. The area overhead of either of the proposed approaches is significantly lower than the area overheads associated with alternate radiation hardening approaches, which commonly require logic duplication or triplication. Note that some designs (such as *frg2*) have a low logic depth and large number of inputs, and consequently, their area overheads are higher.

Table 6.8 reports the total number of gates and the number of hardened gates in a circuit resulting from the use of the proposed circuit tolerant approaches (η^* and η') for both area and delay mapping. In Table 6.8, Columns 2 and 3 report the total number of gates and the number of hardened gates of a radiation-hardened area-mapped design (before resynthesis). Columns 4 and 5 reports these numbers for the radiation-hardened design after resynthesis. Similarly, Columns 5 and 6 report the total number of gates and the number of hardened gates for radiation-hardened delay-mapped designs (before resynthesis), and Columns 7 and 8 report these quantities for radiation-hardened delay-mapped designs after resynthesis.

The delay penalty associated with applying the improved circuit protection approach is presented in Table 6.9. Two different radiation hardened versions (η^* and η') are available for each design and the best among them in terms of area or delay can be selected. In Table 6.9, Column 2 reports the delay (in picoseconds) of a

Table 6.8 Total number of gates and number of hardened gate in different designs

	Area mapping				Delay mapping			
	η^*		η'		η^*		η'	
Ckt	Total no. of gate	No. of hardened gates	Total no. of gate	No. of hardened gates	Total no. of gate	No. of hardened gates	Total no. of gate	No. of hardened gates
alu2	273	45	270	7	429	17	474	7
alu4	537	52	531	11	795	27	845	14
C1355	455	51	450	32	582	32	582	32
C1908	406	45	415	25	579	27	597	25
C3540	893	80	904	22	1,290	46	1,356	24
C499	455	51	450	32	582	32	582	32
C880	308	54	310	26	417	51	445	31
dalu	733	51	747	19	1,064	38	1,082	16
des	2,795	545	2,628	245	3,812	365	4,213	245
frg2	597	221	579	132	846	144	941	137
i2	151	2	151	3	228	3	230	1
i3	110	14	110	6	114	14	118	6
i10	1,775	519	1,787	233	2,507	346	2,792	243

Table 6.9 Delay overhead of the improved circuit protection approach

	Area mapping					Delay mapping				
		Best delay		Best area			Best delay		Best Area	
Ckt	Regular	min(η^*, η')	%Ovh.	min(η^*, η')	%Ovh.	Regular	min(η^*, η')	%Ovh.	min(η^*, η')	%Ovh.
alu2	1,025.881	1,165.1	−3.84	1,214.939	0.27	883.78	1,066.16	1.29	1,073.26	1.96
alu4	1,219.078	1,435.371	2.09	1,533.967	9.1	1,126.59	1,329.84	0.76	1,329.84	0.76
C1355	835.347	984.751	2.58	984.751	2.58	670.35	787.42	1.53	787.42	1.53
C1908	1,206.572	1,385.88	0.67	1,486.142	7.96	1,028.11	1,184.32	1.05	1,184.32	1.05
C3540	1,466.264	1,728.315	2.71	1,772.92	5.36	1,343.16	1,571.99	0.73	1,588.23	1.77
C499	835.347	984.751	2.58	984.751	2.58	670.35	787.42	1.53	787.42	1.53
C880	1,402.478	1,323.711	− 17.58	1,323.711	− 17.58	1,319.60	1,240.00	− 19.69	1,240.00	−19.69
dalu	1,041.554	1,363.747	2.88	1,415.225	6.77	1,008.48	1,232.72	0.93	1,232.72	0.93
des	1,476.792	1,721.303	−20.71	2,595.902	19.57	1,459.22	2,053.42	1.84	2,272.79	12.72
frg2	768.107	930.828	2.23	991.758	8.92	756.23	870.59	− 4.51	957.09	4.97
i2	432.983	477.99	3.42	477.99	3.42	346.60	386.72	2.46	417.15	10.52
i3	160.868	199.782	15.84	233.17	35.2	160.87	194.38	12.71	194.38	12.71
i10	1,879.402	2, 335.109	5.29	2,685.245	21.07	1,882.47	2,315.50	3.09	2,315.50	3.09
AVG			−0.14		8.09			0.29		2.60

regular area-mapped design. Column 3 reports the delay of radiation-hardened area-mapped design with the best delay. Column 4 reports the percentage delay overhead for this design. Column 5 reports the delay of the radiation-hardened area-mapped design with the best area and Column 6 reports the percentage delay overhead for this design. Similarly, Column 7 reports the delay (in picoseconds) of a regular delay-mapped design. Column 8 reports the delay of the radiation-hardened delay-mapped design with the best delay. Column 9 reports the percentage delay overhead. Column 10 reports the delay of the radiation-hardened delay-mapped design with the best area, and Column 11 reports the percentage delay overhead for this design. Note from this table that the circuit-level delay overhead of the improved circuit protection algorithm is as low as 0.29% on average for delay mapped designs, and about −0.14% for area mapped designs.

The placed-and-routed area penalty associated with applying the improved circuit protection approach is presented in Table 6.10. In Table 6.10, Column 2 reports the placed-and-routed area (in μm^2) of a regular area-mapped design. Column 3 reports the area of the radiation-hardened area-mapped circuits with the best delay. Column 4 reports the percentage area overhead for the corresponding design. Column 5 reports the area of the radiation-hardened area-mapped design with the best area and Column 6 reports the percentage area overhead for this design. Similarly, Column 7 reports the area (in μm^2) of a regular delay-mapped design. Column 8 reports the area of the radiation-hardened delay-mapped circuit with the lowest delay. Column 9 reports the percentage area overhead for the corresponding circuit. Column 10 reports the area of the radiation-hardened delay-mapped designs with the least area, and Column 11 reports the percentage area overhead of corresponding design. As reported in Table 6.10, the circuit-level area overhead of the improved circuit protection algorithm is 23.75% on average for delay mapped designs, and about 29.33% for area mapped designs.

6.5 Chapter Summary

In this chapter, a novel circuit design approach was presented for radiation hardened digital electronics. The proposed approach uses shadow gates to protect the primary gate, in case it is struck by radiation. The gate which is to be protected is locally duplicated, and a pair of diode-connected transistors (or diodes) are connected between the outputs of the original and the shadow gates. These transistors (diodes) turn on when the voltages of the two gates deviate during a radiation strike. The delay overhead of the proposed approach per library gate is about 1.76%. The area overhead of this approach is 277% per library gate.

In addition, a variable depth protection approach was also presented to perform *circuit-level radiation hardening with very low delay and area overheads*. In this approach, the number of gates that need to be protected are minimized. The resulting circuit is made radiation hard, with a very low area and delay penalty (28% and 3% on average, for delay mapped designs) compared with an unprotected circuit. In

Table 6.10 Area overhead of the improved circuit protection approach

	Area mapping					Delay mapping				
		Best delay		Best area			Best delay		Best area	
Ckt	Regular	$\min(\eta^*, \eta')$	%Ovh.	$\min(\eta^*, \eta')$	%Ovh.	Regular	$\min(\eta^*, \eta')$	%Ovh.	$\min(\eta^*, \eta')$	%Ovh.
alu2	1,045.88	1,418.28	35.61	1,215.22	16.19	1,397.26	1,569.74	12.34	1,569.74	12.34
alu4	1,994.52	2,470.09	23.84	2,279.11	14.27	2,470.09	2,756.25	11.59	2,756.25	11.59
C1355	1,592.01	1,994.52	25.28	1,994.52	25.28	1,728.9	2,279.11	31.82	2,279.11	31.82
C1908	1,569.74	1,994.52	27.06	1,799.46	14.63	1,799.46	2,225.95	23.7	2,225.95	23.7
C3540	3,183.22	3,916.26	23.03	3,573.65	12.27	4,022.1	4,572.46	13.68	4,515.84	12.28
C499	1,569.74	1,994.52	27.06	1,994.52	27.06	1,728.9	2,279.11	31.82	2,279.11	31.82
C880	1,045.88	1,418.28	35.61	1,418.28	35.61	1,397.26	1,764	26.25	1,764	26.25
dalu	2,470.09	2,996.47	21.31	2,965.89	20.07	3,310.85	3,573.65	7.94	3,573.65	7.94
des	9,964.03	1,6842.85	69.04	13,731.15	37.81	12,139.63	17,800.9	46.63	15,490.29	27.6
frg2	1,994.52	4,201.63	110.66	3,916.26	96.35	2,611.21	4,238.01	62.3	4,147.36	58.83
i2	685.39	730.08	6.52	730.08	6.52	872.61	872.61	0	872.61	0
i3	495.51	670.81	35.38	600.25	21.14	495.51	600.25	21.14	600.25	21.14
i10	6,037.29	12,016.54	99.04	9,304.53	54.12	7,705.33	11,054.42	43.46	11,054.42	43.46
AVG			41.50		29.33			25.59		23.75

practice, however, a very small fraction of gates need to be protected. Another approach was presented, which reduces the area and delay penalty based on the desired cost function. With the improved circuit protection algorithm, radiation tolerant circuits are obtained with a very low area penalty as low as 23.75% and a delay penalty as low as −0.14% on average.

It is possible to use the proposed gate level hardening approach to memory elements, or even the gates that drive memory elements. In this way, the approach presented in this chapter can protect both combinational and sequential circuits from radiation events. The next chapter describes the split-output-based hardening approach.

References

1. R. Garg, N. Jayakumar, S. P. Khatri, and G. Choi, "A design approach for radiation-hard digital electronics," in *Proc. of the Design Automation Conf.*, July 2006, pp. 773–778.
2. R. Garg and S. P Khatri, "A novel, highly SEU tolerant digital circuit design approach," in *Proc. of the Intl. Conf. on Computer Design*, Oct. 2008, pp. 14–20.
3. W. Massengill, M. Alles, and S. Kerns, "SEU error rates in advanced digital CMOS," in *Proc. of the European Conf. on Radiation and Its Effects on Components and Systems*, Sep. 1993, pp. 546 – 553.
4. J. Pickle and J. Blandford, "CMOS RAM cosmic-ray-induced error rate analysis," *IEEE Transactions on Nuclear Science*, vol. NS-29, pp. 3962–3967, 1981.
5. J. Wang, B. Cronquist, and J. McGowan, "Rad-hard/hi-rel FPGA," in *Proc. of the ESA Electronic Components Conf.*, April 1997, pp. 1–4.
6. R. Garg, N. Jayakumar, S. P. Khatri, and G. Choi, "A design approach for radiation-hard digital electronics," in *Proc. of the Design Automation Conf.*, July 2006, pp. 773–778.
7. Q. Zhou and K. Mohanram, "Gate sizing to radiation harden combinational logic," *IEEE Transactions on Computer-Aided Design of Integrated Circuits and Systems*, vol. 25, no. 1, pp. 155–166, Jan. 2006.
8. Q. Zhou and K. Mohanram, "Transistor sizing for radiation hardening," in *Proc. of the Intl. Reliability Physics Symposium*, April 2004, pp. 310–315.
9. T. May and M. Woods, "Alpha-particle-induced soft errors in dynamic memories," *IEEE Transaction on Electron Devices*, vol. ED-26, pp. 2–9, Jan 1979.
10. B. Gill, M. Nicolaidis, F. Wolff, C. Papachristou, and S. Garverick, "An efficient BICS design for SEUs detection and correction in semiconductor memories," in *Proc. of the Conf. on Design Automation and Test in Europe*, March 2005, pp. 592–597.
11. G. Agrawal, L. Massengill, and K. Gulati, "A proposed SEU tolerant dynamic random access memory (DRAM) cell," in *IEEE Transactions on Nuclear Science*, Dec 1994, vol. 41, pp. 2035–2042.
12. M. Caffrey, P. Graham, E. Johnson, and M. Wirthli, "Single-event upsets in SRAM FPGAs," in *Proc. of the Intl. Conf. on Military and Aerospace Programmable Logic Devices*, Sep. 2002, pp. 1–6.
13. C. Carmichael, E. Fuller, M. Caffrey, P. Blain, and H. Bogrow, "SEU mitigation techniques for Virtex FPGAs in space applicaions," in *Proc. of the Intl. Conf. on Military and Aerospace Programmable Logic Devices*, Sep. 1999, pp. 1–8.
14. S. Whitaker, J. Canaris, and K. Liu, "SEU hardened memory cells for a CCSDIS Reed Solomon encoder," *IEEE Transactions on Nuclear Science*, vol. 38, no. 6, pp. 1471–1477, 1991.
15. M. N. Liu and S. Whitaker, "Low power SEU immune CMOS memory circuits," *IEEE Transactions on Nuclear Science*, vol. 36, no. 6, pp. 1679–1684, 1992.

16. J. S. Cable, E. F. Lyons, M. A. Stuber, and M. L. Burgener, "United States patent 6531739: Radiation-hardened silicon-on-insulator CMOS device, and method of making the same," Available: http://www.freepatentsonline.com/6531739.html, Nov. 2003, pp. 1-16.
17. K. Mohanram and N. A. Touba, "Cost-effective approach for reducing soft error failure rate in logic circuits," in *Proc. of the Intl. Test Conf.*, 2003, pp. 893–901.
18. T. Heijmen and A. Nieuwland, "Soft-error rate testing of deep-submicron integrated circuits," in *Proc. of the IEEE European Test Symposium*, 2006, pp. 247–252.
19. C. Nagpal, R. Garg, and S. P. Khatri, "A delay-efficient radiation-hard digital design approach using CWSP elements," in *Proc. of the Conf. on Design Automation and Test in Europe*, March 2008, pp. 354–359.
20. P. E. Dodd and L. W. Massengill, "Basic mechanisms and modeling of single-event upset in digital microelectronics," *IEEE Transactions on Nuclear Science*, vol. 50, no. 3, pp. 583– 602, 2003.
21. J. P. Hayes, I. Polian, and B. Becker, "An analysis framework for transient-error tolerance," in *Proc. of the VLSI Test Symposium*, 2007, pp. 249–255.
22. C. Zhao, S. Dey, and X. Bai, "Soft-spot analysis: Targeting compound noise effects in nanometer circuits," *Design and Test of Computers*, vol. 22, no. 4, pp. 362–375, 2005.
23. C. Zhao and S. Dey, "Improving transient error tolerance of digital VLSI circuits using robustness compiler (ROCO)," in *Proc. of the Intl. Symposium on Quality Electronic Design*, 2006, pp. 133–140.
24. S. Mitra, N. Seifert, M. Zhang, and K. Kim, "Robust system design with built-in soft-error resilience," *IEEE Transactions on Computers*, vol. 38, pp. 43–52, Feb. 2005.
25. T. Fukai, Y. Nakahara, M. Terai, S. Koyama, and Y. Morikuni et al., "A 65 nm-node CMOS technology with highly reliable triple gate oxide suitable for power-considered system-on-a-chip," *Proc. of the Symposium on VLSI Technology*, pp. 83–84, June 2003.
26. Q. Lin, M. Ma, T. Vo, J. Fan, X. Wu, R. Li, and X. Li, "Design-for-manufacture for multi-gate oxide CMOS process," in *Proc. of the Intl. Symposium on Quality Electronic Design*, 2007, pp. 339–343.
27. B. Amelifard, F. Fallah, and M. Pedram, "Reducing the sub-threshold and gate-tunneling leakage of SRAM cells using dual-Vt and dual-Tox assignment," in *Proc. of the Conf. on Design Automation and Test in Europe*, 2006, pp. 995–1000.
28. Altera Inc., San Jose, CA, "Stratix III programmable power," May 2007, pp. 1-12, Available: http://www.altera.com/literature/wp/wp-01006.pdf.
29. L. Nagel, "Spice: A computer program to simulate computer circuits," in *University of California, Berkeley UCB/ERL Memo M520*, May 1995.
30. Y. Cao, T. Sato, D. Sylvester, M. Orshansky, and C. Hu, "New paradigm of predictive MOSFET and interconnect modeling for early circuit design," in *Proc. of IEEE Custom Integrated Circuit Conf.*, June 2000, pp. 201–204.
31. P. McGeer, A. Saldanha, R. Brayton, and A. Sangiovanni-Vincentelli, *Logic Synthesis and Optimization*, chapter Delay Models and Exact Timing Analysis, pp. 167–189, US Kluwer Academic Publishers, Boston, MA, 1993.
32. E. M. Sentovich, K. J. Singh, L. Lavagno, C. Moon, R. Murgai, A. Saldanha, H. Savoj, P. R. Stephan, R. K. Brayton, and A. L. Sangiovanni-Vincentelli, "SIS: A system for sequential circuit synthesis," Tech. Rep. UCB/ERL M92/41, Electronics Research Laboratory, Univ. of California, Berkeley, May 1992.
33. Cadence Design Systems, Inc., San Jose, CA, *Envisia Silicon Ensemble Place-and-route Reference Manuals*, Nov 1999.

Chapter 7
Split-output-based Radiation Tolerant Circuit Design Approach

7.1 Introduction

This chapter presents the second circuit level radiation hardening approach (the split-output-based hardening approach) developed in this monograph. The split-output-based hardening approach exploits the fact that if a gate is implemented using only PMOS (NMOS) transistors, then a radiation particle strike can result only in logic 0–1 (1–0) flip. On the basis of this observation, radiation hardened variants of regular static CMOS gates are derived. In particular, the PMOS and NMOS devices of regular gates are separated, and the gate output is split into two signals. One of these outputs of the radiation tolerant gate is generated using PMOS transistors, and it drives other PMOS transistors (only) in its fanout. Similarly, the other output (generated from NMOS transistors) only drives other NMOS transistors in its fanout. Now, if a radiation particle strikes one of the outputs of the radiation tolerant gate, then the gates in the fanout enter a high-impedance state, and hence preserve their output values.

Split-output-based radiation hardened gates exhibit an extremely high degree of radiation tolerance, which is validated at the circuit level. Using these gates, circuit level hardening is performed based on logical masking, to selectively harden those gates in a circuit which contribute maximally to the soft error failure of the circuit. The gates whose outputs have a low probability of being logically masked are replaced by their radiation tolerant counterparts, such that the digital design achieves a soft error rate reduction of a specified amount (typically 90%). Experimental results demonstrate that this reduction is achieved with a modest layout area and delay penalty.

In the remainder of this chapter, Sect. 7.2 briefly discusses some additional previous work (in addition to the previous work presented in Chap. 6) on radiation hardening of circuits. Section 7.3 describes the split-output clamping-based radiation tolerant combinational circuit design approach. Experimental results are presented in Sect. 7.4, while the chapter summary is provided in Sect. 7.5.

R. Garg and S. P. Khatri, *Analysis and Design of Resilient VLSI Circuits: Mitigating Soft Errors and Process Variations*, DOI 10.1007/978-1-4419-0931-2_7,

7.2 Related Previous Work

In addition to the existing circuit level hardening approaches discussed in Sect. 6.2, some other approaches [1, 2, 3] use the fact that a particle hit induces a current which always flows from the n-type diffusion to the p-type diffusion, through a p-n junction. This means that if a flip-flop is made up of only PMOS (NMOS) transistors, then a radiation particle strike cannot flip the node voltage from 1 to 0 (0–1). The authors of [1] use this observation to design a radiation hardened flip-flop (with two inputs and two outputs), by separating the NMOS and the PMOS transistors in the flip-flop. However, their flip-flop design has significantly higher leakage currents, since some nodes have nonrail voltages in steady state. The authors of [2] alleviate this problem by adding a few more transistors to the radiation tolerant flip-flop design of [1].

In [3], the author borrows the idea of [1] to design a radiation tolerant standard cell library. However, these hardened cells have significantly larger leakage currents due to nonrail voltage levels at the output nodes of the gates. This is a significant problem because leakage power in today's technologies is comparable to or greater than switching power [4]. Further, the author of [3] did not describe a methodology to implement a radiation tolerant circuit using the radiation tolerant standard cell library, and hence did not report the area and delay overheads of the resulting radiation tolerant circuit. Also, the transistor of radiation tolerant standard cells of [3] have to be sized very carefully to allow correct operation. In contrast, the radiation tolerant standard cells proposed in this chapter do not suffer from these issues. This is described in Sect. 7.3.1.

7.3 Proposed Split-output-based Radiation Hardening

In Sect. 7.3.1, the radiation-tolerant standard cell design approach proposed in this chapter is described, along with an explanation of how these hardened gates are derived from regular gates. The circuit level hardening approach is described in Sect. 7.3.2.1. For circuit hardening, only those gates in a circuit which contribute maximally toward the soft error failure rate of the circuit are replaced by their hardened counterparts. The circuit level hardening approach presented in this chapter achieves a soft error failure rate reduction of an order of magnitude (i.e., 90% reduction in the soft error rate) for several ISCAS and MCNC benchmark circuits. Section 7.3.3 presents an analysis to estimate the critical charge for the hardened circuit obtained by using the approach presented in this chapter.

7.3.1 Radiation Tolerant Standard Cell Design

As mentioned in Sect. 7.2, a radiation particle strike induces a current, which always flows from the n-type diffusion to the p-type diffusion through a p-n junction [1].

This implies that if a gate is made up of only PMOS (NMOS) transistors then a radiation particle strike cannot flip the node voltage from 1 to 0 (0–1). In other words, if a particle strikes the diffusion of a PMOS transistor of an inverter (constructed exclusively using PMOS devices) whose output is at logic 1, then this particle strike will not cause the output node voltage to experience an SET. Similarly, a particle strike at the diffusion of a NMOS transistor of the inverter constructed exclusively using NMOS devices (with an output node at logic 0) will not result in an SET. This is a key idea, which conveys that if a logic circuit is made up of only PMOS (NMOS) transistors, then that logic circuit will be tolerant to node voltage changes from logic values 1 to 0 (0–1). This concept is used in the hardening approach presented in this chapter to design radiation tolerant standard cells.

Consider two regular inverters as shown in Fig. 7.1a. Radiation particle strikes at M1 and M2 of INV1 can result in both positive or negative voltage glitches at the *out*1 node (since the PMOS and NMOS transistors are both connected to the *out*1 node). The voltage glitch at the *out*1 node can affect the node voltage of *out*2, which can lead to a soft error. To avoid an error event due to radiation particle strikes at the diffusions of M1 or M2, INV1 needs to be hardened. The steps to harden INV1 are as follows.

First, M1 (NMOS) and M2 (PMOS) of INV1 in Fig. 7.1a are separated from each other, and the resulting circuit is shown in Fig. 7.1b. The inverter INV1 shown in Fig. 7.1b has 2 inputs (*inp* and *inn*) and 2 outputs (*out*1*p* and *out*1*n*). Both the inputs and both the outputs of INV1 have the same polarity. Note that the output nodes *out*1*p* and *out*1*n* of INV1 drive only PMOS and NMOS transistors of the gates in their fanout, respectively (*out*1*p* drives M4 of INV2 and *out*1*n* drive M3 of INV2 in Fig. 7.1b). Also, note that the inverter INV2 is also modified such that two different input signals (of the same polarity) drive the transistors M3 and M4. In this chapter, an *n* input regular gate (such as the inverter INV2 of Fig. 7.1b),

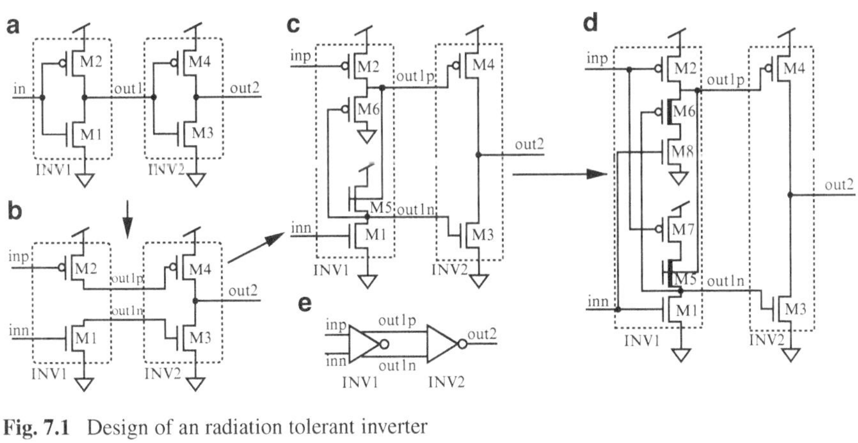

Fig. 7.1 Design of an radiation tolerant inverter

whose inputs to PMOS and NMOS transistors are separated, is referred to as the *modified* regular gate. Note that such a gate has $2n$ inputs. Also note that a n input radiation-hardened gate has $2n$ inputs and 2 outputs.

However, in the INV1 circuit of Fig. 7.1b, $out1p$ ($out1n$) can only charge (discharge) to 1 (0). To get the opposite transitions at node $out1p$ and $out1n$, two additional transistors M5 (NMOS) and M6 (PMOS) are added, and connected as shown in Fig. 7.1c. The inverter INV1 of Fig. 7.1c works as follows. Assume that both *inn* and *inp* are at a logic 0 value, and $out1p$ and $out1n$ are at logic 1. Now assume that both *inn* and *inp* transition to logic 1 due to which transistor M1 turns on and M2 turn off. The turning on of M1 pulls the node $out1n$ down to logic 0, which then turns on M6. Since M6 is ON, $out1p$ drives a weak logic 0[1]. Both the $out1p$ and $out1n$ nodes are now at logic 0, because of which the output of INV2 (*out*2) switches to logic 1. Now when both inputs of INV1 (*inn* and *inp*) change to logic 0, then transistor M1 turns off and M2 turns on. As M2 is on, $out1p$ charges to logic 1, which turns M5 on and hence, node $out1n$ is pulled to a weak logic 1 (VDD- V_T^{M5} V). Thus, the circuit INV1 of Fig. 7.1c behaves like an inverter, with the output node $out1p$ ($out1n$) switching between VDD and $|V_T^{M6}|$ (VDD - V_T^{M5} and GND). Note that the inverter INV1 of Fig. 7.1c has a high leakage power dissipation because nodes $out1p$ and $out1n$ switch between nonrail voltage values. Specifically, when *inn* and *inp* are at GND, then $out1p$ is at VDD and $out1n$ is at VDD-V_T^{M5}. Because of this, M6 is not fully turned off while M2 is completely on. Hence, there is a static power dissipation through M2 and M6. Similarly, when *inn* and *inp* are at the VDD value, there is a static power dissipation through M5 and M1. Also note that INV1 of Fig. 7.1c is tolerant to radiation particle strikes at $out1p$ and $out1n$. This will be explained shortly, after discussing a modification to the proposed INV1 design of Fig. 7.1c, which is not only tolerant to a radiation particle strike, but also significantly reduces the static power dissipation. Experimental results show that this modification yields a reduction in leakage of 2 orders of magnitude.

To reduce the static power dissipation in INV1 of Fig. 7.1c, two more transistors (M7 and M8) are added to INV1 and the resulting inverter is shown in Fig. 7.1d. Now when the inputs of INV1 (*inn* and *inp*) of Fig. 7.1d are at the GND value, again the $out1n$ node is at VDD-V_T^{M5} due to which M6 is still not fully turned off. However, M8 is completely off (since *inn* is at the GND value) and hence there is no static power dissipation through M2, M6, and M8. Similarly, there is no static power dissipation through M7, M5, and M1 when the inputs *inn* and *inp* are at the VDD value. Note that the transistors M5 and M6 of INV1 in Fig. 7.1d are selected to be low threshold voltage transistors (indicated by a thicker line in the figure). This is done so as to increase the voltage swing at nodes $out1p$ and $out1n$, and bring them closer to the rail voltages. Note that the reduced voltage swings at $out1p$ and $out1n$ do not increase the leakage currents in INV2 of Fig. 7.1d. This is because when the node $out1p$ is at the $|V_T^{M6}|$ value, then $out1n$ is at the GND value, because of which M3 is completely turned off while M4 is turned on. Similarly, when $out1p$ is at the

[1] The node $out1p$ falls to $|V_T^{M6}|$ V. Note that V_T^{M6} is the threshold voltage of PMOS transistor M6.

VDD value then *out*1*n* is at VDD-V_T^{M5}, and hence M3 is turned on while M4 is completely turned off. Therefore, the leakage currents in INV2 do not increase due to nonrail voltage swing at its inputs.

The inverter INV1 of Fig. 7.1d is tolerant to a radiation strike at *out*1*p* and *out*1*n*. Consider the case when the nodes *inp* and *inn* are at VDD, which implies that *out*1*p* and *out*1*n* are at $|V_T^{M6}|$ and GND, respectively, and *out*2 is at the VDD value. Now suppose a radiation particle strikes at node *out*1*p* (the radiation particle strikes either M2 or M6), which increases the voltage at node *out*1*p* to VDD (because of the positive charge collection at *out*1*p*). Because of this, M4 of INV2 turns off and M5 turns on. However, the node *out*1*n* remains at GND value because M7 is in cutoff. Therefore, M3 also remains off. Thus, the node *out*2 remains at the VDD value in a high impedance state. Eventually, the charge collected at *out*1*p* dissipates through M6 and M8 (since *inp* and *inn* are at VDD), which brings the voltage at *out*1*p* node back to $|V_T^{M6}|$. At this point, M4 turns on again. In this way, a radiation strike at *out*1*p* does not affect the node voltage of *out*2. Similarly, a particle strike at *out*1*n* does not affect the node *out*2 when *inn* and *inp* are at the GND value.

To summarize, a radiation particle at *out*1*p* (*out*1*n*) node can only result in a positive (negative) glitch, since only PMOS (NMOS) transistors are connected to it. Also this positive (negative) glitch at *out*1*p* (*out*1*n*) does not propagate to *out*2. This is because the *out*1*p* (*out*1*n*) node drives only the PMOS (NMOS) transistor of INV2, which goes into cutoff mode when a positive (negative) glitch appears at *out*1*p* (*out*1*n*) node. A radiation particle strike at M8 can be of any significance only when *out*1*p* is at the VDD value (since a radiation particle strike at the NMOS transistor can only result in a negative glitch). However, when *out*1*p* is at VDD, M6 is turned off and hence a particle strike at M8 does not affect the node voltage of *out*1*p*. Similarly, a radiation particle strike at M7 does not affect the voltage of the *out*1*n* node. In this way, INV1 of Fig. 7.1d is tolerant to radiation particle strikes since a particle strike at either of its output nodes does not affect the output of its fanout gates (*out*2 of INV2 in Fig. 7.1). To validate this analysis, inverters INV1 and INV2 of Fig. 7.1d were implemented using a 65 nm PTM [5] model card with VDD = 1.0 V. Radiation particle strikes were simulated at *out*1*p* (at time = 0.8 ns) and *out*1*n* (at time = 2.4 ns) nodes, with $Q = 150$ fC, $\tau_\alpha = 150$ ps and $\tau_\beta = 38$ ps. The values of Q, τ_α and τ_β were obtained from [6]. The voltage waveforms at *out*1*p*, *out*1*n*, and *out*2 are shown in Fig. 7.2.

Note that the inverter INV1 of Fig. 7.1c is also tolerant to radiation particle strikes in a similar manner as INV1 of Fig. 7.1d. The radiation tolerant standard cell library designed by the author of [3] is similar to the inverter INV1 of Fig. 7.1c. However, there are many issues associated with it. First, there is a static power dissipation in the INV1 of Fig. 7.1c as described earlier. Second, the transistor M2 and M6 (M1 and M5) of INV1 have to be sized very carefully to allow for correct inverter operation. The transistor M1 (M2) is sized larger than M5 (M6). Last, when a radiation particle strike at *out*1*p* results in a positive glitch at *out*1*p*, M5 turns on and hence, the voltage at node *out*1*n* is determined by the relative drive strength of M1 and M5. Therefore, a small positive glitch can occur at *out*1*n* (as M1 is sized larger than M5), which can turn on M3 for long enough to pull the node *out*2 low. In contrast to

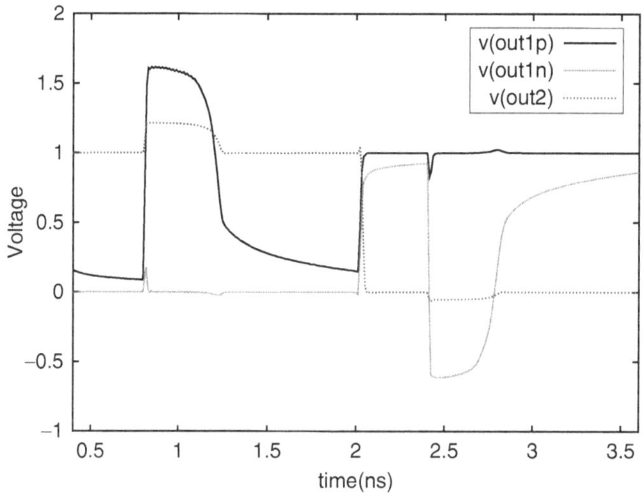

Fig. 7.2 Radiation particle strike at *out*1*p* and *out*1*n* of INV1 of Fig. 7.1d

the radiation tolerant inverter design of [3] (or the INV1 of Fig. 7.1c), the radiation tolerant inverter design, which is proposed in this chapter (shown in Fig. 7.1d), does not suffer from these issues. The leakage currents of INV1 of Fig. 7.1c and INV1 of Fig. 7.1d were also extracted through SPICE simulations. The leakage currents of INV1 of Fig. 7.1c are 0.939 μA and 1.936 μA when both inputs (*inp* and *inn*) are at logic 0 and 1, respectively. In contrast, the leakage currents of the radiation hardened inverter INV1 of Fig. 7.1d are 12.7 nA and 13.5 nA (for logic 0 and 1 at both inputs). Therefore, as mentioned earlier, the radiation hardened gates proposed in this chapter have 2 orders of magnitude lower leakage compared with the hardened gates of [3]. Figure 7.1e also shows the symbolic diagram of the radiation tolerant inverter (INV1) and the *modified* regular inverter (INV2) of Fig. 7.1d.

The radiation hardening approach described above can be applied to any static CMOS gate, including complex gates. Figure 7.3 shows radiation tolerant NAND2, NAND3, NOR2, and NOR3 gates designed using this approach. Figure 7.4 shows the *modified* regular NAND2, NAND3, NOR2, NOR3 gates. As shown in Fig. 7.3a, the radiation tolerant 2-input NAND (NAND2) gate has a total of 4 inputs and 2 outputs. The inputs *in*1*p* and *in*1*n* (*in*2*p* and *in*2*n*) correspond to the first input *in*1 (second input *in*2) of a regular 2-input NAND gate. The two outputs *out*1*p* and *out*1*n* of the radiation tolerant 2-input NAND gate of Fig. 7.3 drive the PMOS and the NMOS transistors of the gates in its fanout. In general, an n-input static CMOS gate requires $4n + 2$ transistors when implemented using the radiation hardening approach proposed in this chapter, in contrast to $2n$ transistors for its regular static CMOS counterpart.

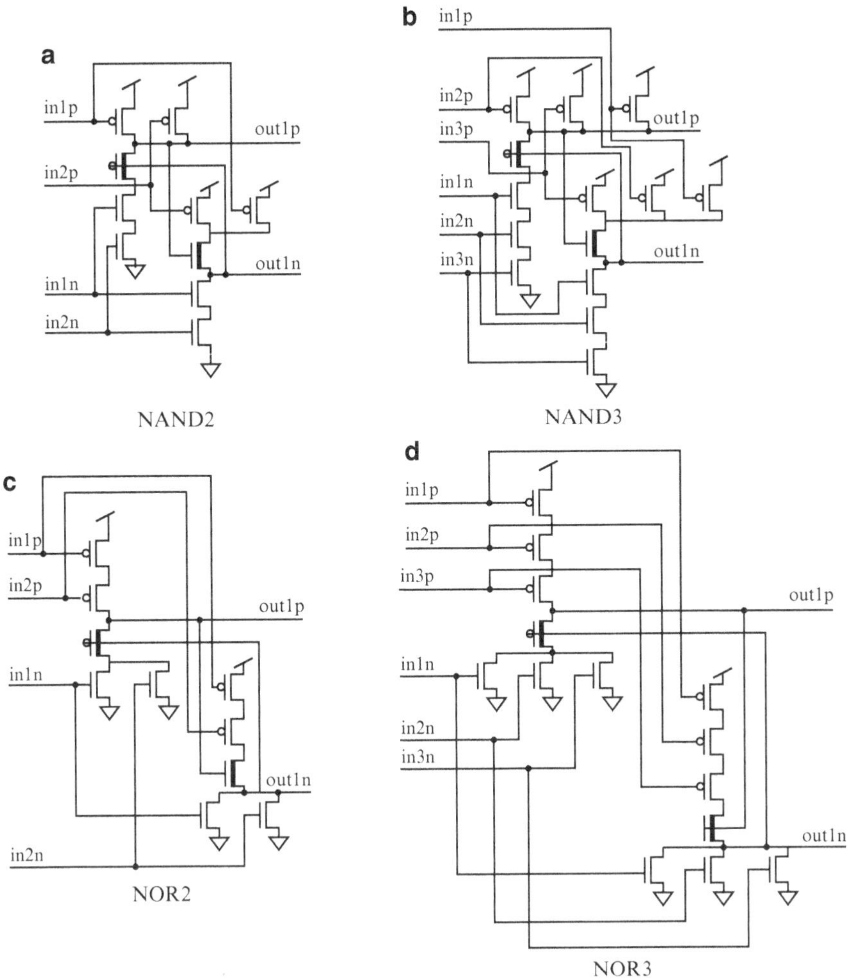

Fig. 7.3 Radiation tolerant gates

7.3.2 *Circuit Level Radiation Hardening*

To keep the area and delay overhead low, selective hardening needs to be performed by only protecting those gates in a circuit that have a significant contribution to the soft error failure rate of the circuit. Whether a voltage glitch induced by a radiation particle strike at any gate in a circuit that propagates to the primary outputs and results in a failure depends upon three masking factors: logical, electrical, and temporal masking, as described in Sect. 1.1.

The sensitive gates in a circuit are those gates that have small values for these masking factors, and hence these gates contribute significantly to the soft error

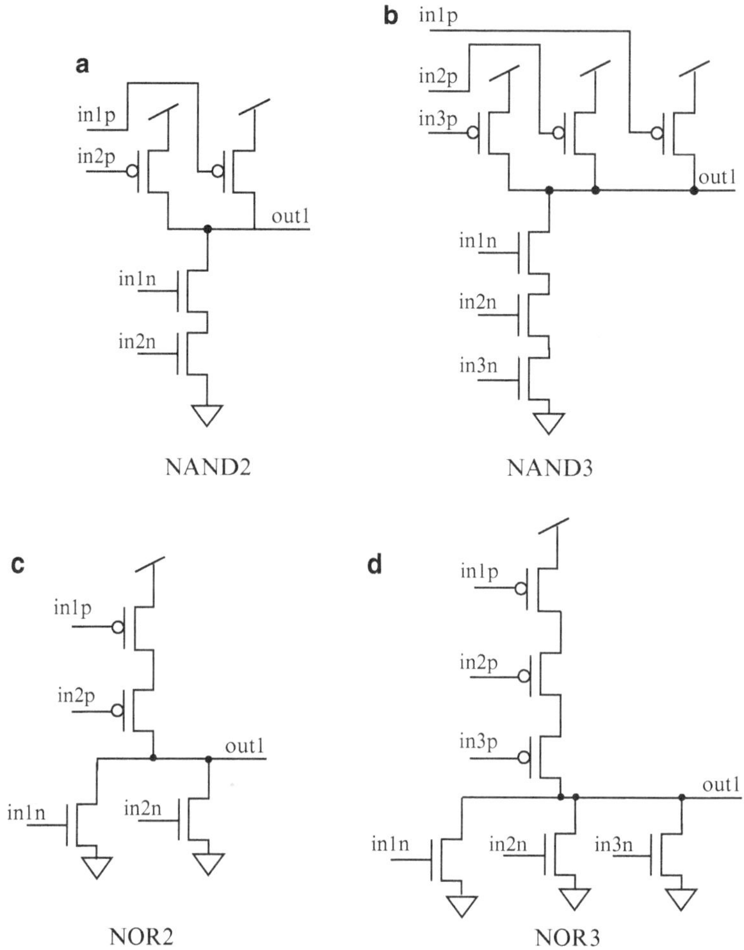

Fig. 7.4 *Modified* regular gates

failure of the circuit. These are the gates in a circuit which are needed to be protected by replacing them with the hardened gates presented earlier, to significantly improve the radiation tolerance of the circuit. The split-output-based hardening approach uses logical masking to identify such sensitive gates in a circuit. The approach used to identify these gates is described next.

7.3.2.1 Identifying and Protecting Sensitive Gates in a Circuit

To identify the sensitive gates in a circuit, a measure of the logical masking at all gates in the circuit is computed. The logical masking at a gate is computed as the probability of the absence of a functionally sensitizable path from the gate to any primary output of the circuit. The computation of the probability of logical masking at a gate is carried out in the same manner as proposed in [6], as described later in this section. As mentioned in [6], the probability of logical masking ($P_{\rm LM}$) at a gate G is $1 - P^{\rm G}_{\rm Sen}$, where $P^{\rm G}_{\rm Sen}$ is the probability of sensitization of gate G. The probability of sensitization is defined as the probability of the existence of at least one functionally sensitizable path from the gate G to any primary output of the circuit.

To calculate the probability of sensitization $P_{\rm Sen}$, N random vectors were applied to primary inputs of a circuit. For each vector, fault simulation was performed on all gates in the circuit to determine whether the fault is sensitized and observable at one or more primary output. For a gate G, the number of vectors ($S_{\rm G}$) that were able to sensitize any fault (both G-stuck-at-1 and G-stuck-at-0) at G to the primary output(s) is recorded. Note that $S_{\rm G}$ is the summation of the number of vectors, which simulate the fault at G (when the output of G is at logic 0 or logic 1). Now the sensitization probability for the gate G ($P^{\rm G}_{\rm Sen}$) is calculated as $S_{\rm G}/{\rm N}$. The value of N used was 10,000. A gate which has high probability of sensitization is a sensitive gate which needs to be protected.

After computing the sensitization probabilities (or logical masking probabilities) for all the gates in the circuit (η), the sensitive gates are identified and protected using Algorithm 3. For a given circuit η, all gates $G \in \eta$ are sorted in a decreasing order of their sensitization probabilities, and stored in a list *LIST*. Then the top K gates in the *LIST* are protected (by replacing them with the hardened gates designed in Sect. 7.3.1) so that the required tolerance against radiation particle strikes is achieved. The resulting hardened circuit is referred as η^*.

In this work, a circuit is called protected when the soft error rate reduces by an order of magnitude. To achieve this, it is required to protect gates in list *LIST* (in decreasing order of sensitization probability) until 90% *coverage* is achieved. The *coverage* is defined as [6].

$$\text{Coverage} = \frac{\sum\limits_{\forall \text{ hardened gate } G \in \eta^*} P^G_{\rm Sen}}{\sum\limits_{\forall \text{ gate } G \in \eta} P^G_{\rm Sen}} \times 100. \tag{7.1}$$

Algorithm 3 Radiation Hardening for a Circuit η

Harden_circuit(η)
$LIST = sort(G \in \eta, P^G_{Sen})$
$i = 0$
while *required tolerance to radiation is not achieved* **do**
  $G = LIST(i)$
  Replace G *by* $G_{hardened}$
  $i = i + 1$
end while
return η^*

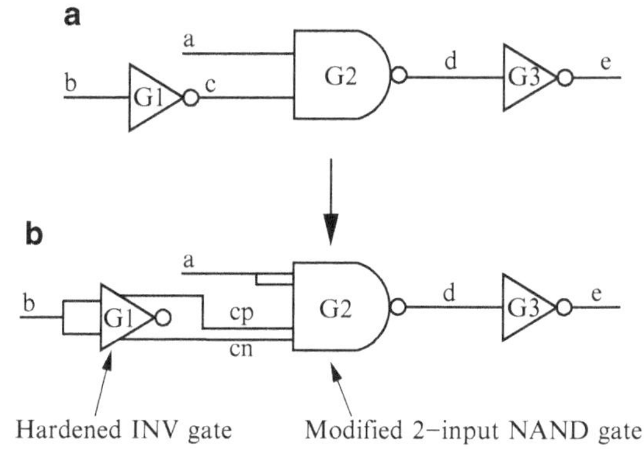

Fig. 7.5 Part of a circuit

It was demonstrated in [6] that *coverage* is a good estimate for soft error failure rate reduction. Ninety percent coverage corresponds to an order of magnitude reduction in the soft error rate. To achieve 90% coverage, K gates have to be protected using Algorithm 3. Note that a gate G is protected by replacing it with its hardened version, which is obtained by the gate hardening technique as described in Sect. 7.3.1. For example, consider a circuit fragment shown in Fig. 7.5a. Note that all the gates in Fig. 7.5a are regular gates. Suppose that the gate $G1$ has a very high sensitization probability and it needs to be protected such that a radiation particle strike at its output should not affect the gates in its fanout ($G2$). To achieve this, the gate $G1$ of Fig. 7.5a is replaced with the hardened inverter gate of Fig. 7.1d. The resulting circuit is shown in Fig. 7.5b. While replacing the gate $G1$ with its hardened version, it is also required to replace all the regular gates in its fanout ($G2$) with their *modified* regular gates,[2] because a hardened gate has two outputs (one output drives only PMOS transistors of the gates in its fanout, and the second drives only the NMOS transistors of the fanout gates of $G1$). Therefore, $G2$ in Fig. 7.5a is also replaced by its *modified* regular version in Fig. 7.5b. To verify that this replacement strategy works, the circuit shown in Fig. 7.5b was implemented using a 65 nm PTM [5] model card, and radiation particle strikes were simulated at nodes cp (at time = 0.8 ns) and cn (at time = 2.4 ns) with $Q = 150$ fC, $\tau_\alpha = 150$ ps, and $\tau_\beta = 38$ ps. The waveform at nodes cp, cn, and d are shown in Fig. 7.6. Figure 7.6 shows that the radiation particle strikes at cp and cn do not have any detrimental effects on the voltage at node d. Hence, the hardening technique proposed in this chapter is able to selectively harden $G1$ of Fig. 7.5a.

The gates at the primary output of a circuit are always sensitive and have a sensitization probability equal to 1. Therefore, these gates are always replaced by their hardened counterparts. However, the replacement of a regular gate (which has one output) with its hardened counterpart results in two outputs. Therefore, the two outputs of each hardened gate that drives the primary outputs of the circuit now

[2] As mentioned earlier, a *modified* regular n-input gate is same as the regular n-input gate with its inputs to the PMOS and the NMOS transistor disconnected from each other resulting in total of $2n$-inputs. For example, INV2 of Fig. 7.1d is a *modified* regular inverter. The gate of Fig. 7.4a is a *modified* regular NAND2 gate.

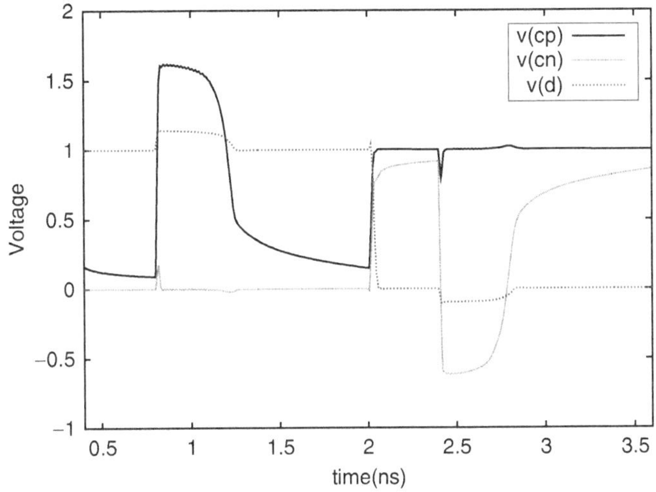

Fig. 7.6 Waveforms at nodes cp, cn, and d of Fig. 7.5b

need to drive a flip-flop (which samples the primary output values). In the proposed approach, the radiation tolerant flip-flop design proposed in [2] (shown in Fig. 7.7) is utilized. This flip-flop design is widely used to implement radiation tolerant VLSI circuits. The radiation tolerant flip-flop of [2] has dual inputs, which correspond to the input D of the regular flip-flop. One of the 2 inputs of the radiation tolerant flip-flop only drives PMOS transistors, and the other input drives only NMOS transistors. Therefore, the hardened gates designed in Sect. 7.3.1 are compatible with radiation tolerant flip-flop of [2].

7.3.3 Critical Charge for Radiation Hardened Circuits

The waveforms shown in Figs. 7.2 and 7.6 suggest that even a large amount of charge dumped by a radiation strike at the output of the proposed hardened gate will not affect the fanout gates' output. Therefore, the approach of Sect. 7.3.2 for circuit level radiation hardening provides 90% coverage against radiation particle strikes from very high energy radiation particles. However, the frequency of circuit operation imposes a limit on the magnitude of the charge dump that can be tolerated by a hardened circuit implemented using the hardening approach proposed in this chapter. This is explained next.

Consider a portion of a hardened circuit as shown in Fig. 7.8a. The waveform of the various nodes, along with *CLK*, are shown in Fig. 7.8b. In Fig. 7.8b, dark lines correspond to the normal operation (no radiation particle strike). The clock period of the hardened circuit is T and the propagation delay of INV2 is d. Assume $out1p$ and $out1n$ switch from logic low to high at $t1$. Now assume that a high energy

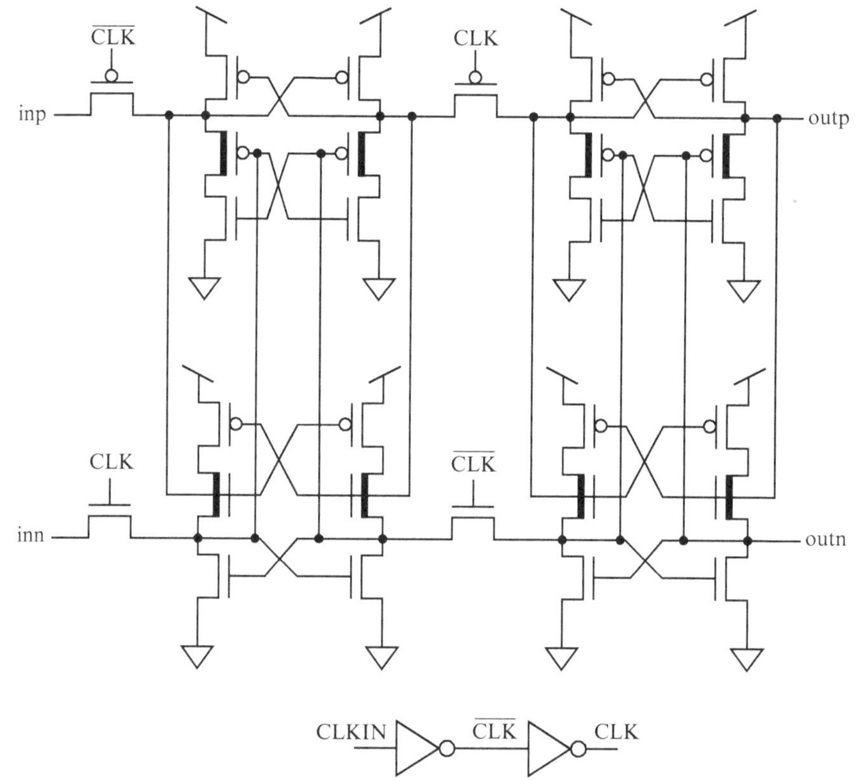

Fig. 7.7 Radiation tolerant flip-flop

radiation particle strikes the *out*1*p* node sometime before $t1$. The particle induces a voltage glitch with the pulse width greater than T and the voltage glitch rises before $t1$ and falls after $T + t1$. As the node *out*1*p* switches to logic 1 before $t1$ (while *out*1*n* is at logic 0); therefore, the node *out*2 enters the high impedance state. At time $t1$, *out*1*n* also switches high (since *in* switches to a low value). Now *out*2 comes out of the high impedance state and switches to logic 0 at the same time as in normal operation. Now at time $T+t1$, *out*1*n* switches to logic 0, and hence the *out*2 node again enters a high impedance state (since *out*1*p* is still at logic 1 due to the radiation strike). When *out*1*p* fall to logic 0 then *out*2 switches to logic 1 as shown in Fig. 7.8b. However, note that the rising *out*2 transition is delayed in comparison to the normal operation. Because of this, the primary output computation may get delayed, potentially resulting in a circuit failure. If the voltage glitch at *out*1*p* had fallen on or before $T+t1$, then the *out*2 node would have switched at the same time as in normal operation, and hence no circuit failure would have been encountered. Thus, the pulse width of the voltage glitch induced by a radiation particle strike at *out*1*p* should be less than the clock period T. Hence the critical charge (Q_{cri}) for the circuit is the maximum amount of charge deposited by a radiation particle such

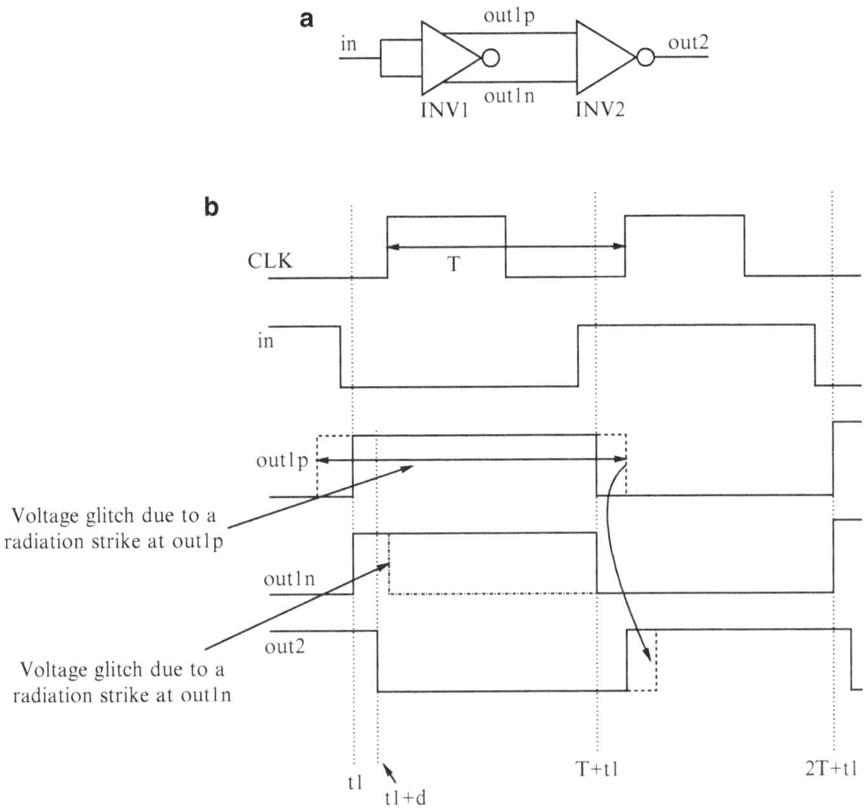

Fig. 7.8 (**a**) Circuit under consideration, (**b**) Waveform at different nodes

that a voltage glitch of pulse width T is encountered in the circuit. As reported in Sect. 7.4, a very large amount of charge should be dumped by a radiation particle to generate a voltage glitch with the pulse width equal to the clock period of a design. This experiment was conducted for the smallest (most sensitive to radiation) gate in the library used in the work presented in this chapter. This is quantified in Sect. 7.4. It was found that the proposed approach is extremely robust to radiation strikes.

Now consider a radiation particle strike just after $t1 + d$, at node $out1n$. Because of the particle strike, $out1n$ switches to logic 0 at $t1 + d$, $out2$ enters the high impedance state with the correct logic value of 0. Even if the pulse width of the negative voltage glitch at $out1n$ is greater than T, it is of no consequence to the node voltage of $out2$. This is because at time $T + t1$ $out1p$ switches to logic 0, hence $out2$ switches to logic 1 at the same time as in normal operation. However, if a radiation particle strikes the $out1n$ node between $t1$ and $t1+d$, then $out2$ enters the high impedance state with the wrong logic value of 1 (since $out1n$ switched to logic 0 before the $out2$ node switches to logic 1). Hence, $out1n$ is vulnerable to a radiation strike when the gates (whose one of the input is $out1n$) is switching their outputs.

To summarize, the maximum tolerable radiation-induced glitch width for the proposed approach is T. Also, the hardened gates designed in this chapter are vulnerable to radiation particle strikes during the time when their fanouts are computing their outputs. For the circuit shown in Fig. 7.8a, INV1 is vulnerable to radiation particle strikes only between $t1$ and $t1 + d$. However, the probability of a particle to strike the $out1n$ node during this time interval is very low[3] and hence, it does not have any impact on the reduction of the overall soft error rate obtained by the proposed approach.

The hardening approach proposed in this monograph can tolerate radiation-induced voltage glitches with a maximum width of T, the clock period of the design. Hence, the proposed hardening approach can provide tolerance against radiation particles of very high energy. In the experimental section, the critical charge (Q_{cri}) values for various benchmark circuits that are hardened using the proposed approach is reported, to support this claim.

7.4 Experimental Results

The performance of the circuit hardening approach proposed in this chapter was evaluated by applying it to several ISCAS and MCNC benchmark circuits. A standard cell library (*LIB*) was implemented using 65 nm PTM [5] model cards, with VDD = 1.0 V. The library (*LIB*) consists of regular INV2X, INV4X, NAND2, NAND3, NOR2, and NOR3 gates. The *modified* regular versions, as well as the hardened versions of all the regular gates in the library *LIB* were also designed. The layouts were created for all these gates using CADENCE SEDSM [9] tools. Several ISCAS and MCNC benchmark circuits were mapped using *LIB*, for both area and delay optimality. From a mapped design, first the sensitization probability of all the gates in the design was computed. Then the sensitive gates in the design were selectively hardened (to achieve a 90% reduction in soft error rate) based on their sensitization probability, using Algorithm 3. The area and the delay results of the regular (unhardened) and the hardened circuits are reported in Tables 7.1 and 7.2.

Table 7.1 reports the layout area results for several benchmark circuits, which are mapped for both area and delay optimality. Note that the layout area for a design was computed by adding the layout area of all the gates in the circuit. Column 1 reports the circuit under consideration. Columns 2 and 3 report the area (in μm^2) of the regular and hardened (obtained using the approach proposed in this chapter) area mapped designs, respectively. Column 7 (8) report the area of regular (hardened) delay mapped designs. Column 4 (9) reports the percentage area overhead for the

[3] As per [7, 8], the maximum solar proton fluence for particles of energy > 1 MeV based on the JPL- 1991 model is $2.91 \times 10^{11}/\mathrm{cm}^2/year$ with 99% confidence. The maximum area of a hardened gate in the library used in this work is $7.69 \times 10^{-8}\mathrm{cm}^2$ and the maximum delay of any gate is 70 ps. Using these values, it can be shown that the probability of a radiation particle to strike $out1n$ between $t1$ and $t1 + d$ is 4.96×10^{-14}.

Table 7.1 Area overheads of the radiation hardened design approach proposed in this chapter

	Area map					Delay map				
Ckt	Regular (μm^2)	Hardened (μm^2)	%Ovh.	Total no. of gates	No. of hard	Regular (μm^2)	Hardened (μm^2)	%Ovh.	Total no. of gates	No. of hard
alu2	667.89	1, 080.92	61.84	303	187	740.39	1, 160.24	56.71	398	218
apex7	417.43	699.96	67.68	214	148	465.76	748.96	60.80	263	163
C1355	949.54	1, 627.32	71.38	516	377	1,015.89	1,699.16	67.26	566	385
C1908	908.68	1,486.05	63.54	479	310	1,020.73	1,624.68	59.17	572	347
C3540	2,177.23	3,312.64	52.15	1,079	588	2,401.32	3,571.66	48.74	1,271	639
C432	348.88	609.23	74.62	181	134	402.93	684.80	69.96	219	155
C499	974.59	1,634.13	67.67	523	362	1,069.50	1,756.28	64.22	597	387
C880	772.90	1,293.15	67.31	389	267	828.71	1,361.26	64.26	457	294
dalu	1,569.98	2,458.44	56.59	781	444	1,799.78	2,822.27	56.81	969	545
alu4	4,093.89	5,945.08	45.22	1,799	782	4, 543.40	6,221.46	36.93	2,469	880
frg2	1,453.54	2,302.90	58.43	723	422	1,768.15	2,736.36	54.76	988	571
AVG			62.40					58.15		

Table 7.2 Delay overheads and Q_{cri} of the proposed radiation hardened design approach

	Area map				Delay map			
	Regular	Hardened			Regular	Hardened		
Ckt	(ps)	(ps)	%Ovh.	Q_{cri}(fC)	(ps)	(ps)	%Ovh.	Q_{cri}(fC)
alu2	1,068.28	1,309.45	22.58	>650	893.62	1,129.06	26.35	>650
apex7	495.00	636.84	28.65	520	451.95	565.13	25.04	330
C1355	636.95	830.21	30.34	>650	639.86	799.49	24.95	>650
C1908	924.56	1,206.91	30.54	>650	926.47	1,205.01	30.06	>650
C3540	1,217.71	1,582.78	29.98	>650	1,139.45	1,530.20	34.29	>650
C432	856.80	1,120.31	30.76	>650	839.02	1,094.36	30.43	>650
C499	670.97	868.22	29.40	>650	655.22	784.30	19.70	>650
C880	923.22	1,157.03	25.32	>650	879.10	1,069.67	21.68	>650
dalu	909.35	1,241.81	36.56	>650	821.68	1,157.31	40.85	>650
alu4	679.15	818.10	20.46	>650	625.48	751.79	20.20	>650
frg2	679.32	905.38	33.28	>650	818.30	1,098.05	34.19	>650
AVG			28.90				27.98	

radiation-hardened area (delay) mapped designs. Column 5 (10) reports the number of gates in the area (delay) mapped designs. The number of hardened gates in the area (delay) mapped designs is reported in Column 6 (11). Observe from Table 7.1 that the average area overhead for the proposed hardening approach is 62.4% and 58.15%, for area and delay mapped designs, respectively. Also, on average, the ratio of the number of hardened gates and the total number of gates in the area and delay mapped designs is 63.1% and 58.7%, respectively.

The delay penalty associated with applying the proposed radiation hardening approach is presented in Table 7.2. Note that the delay for a design reported in Table 7.2 is the summation of the combinational circuit delay (D), the setup time (T_{su}) of the flip-flop and the clock to output (T_{cq}) delay of the flip-flop. Therefore, Table 7.2 reports the clock period ($T = D + T_{su} + T_{cq}$) of a design. The delay of a regular design is obtained by using a static timing analysis tool. The static timing analysis tool was modified to compute the delay of the radiation hardened designs. First, all hardened gates were characterized to construct two-dimensional pin-to-output delay lookup tables for different load values on the two outputs ($outp$ and $outn$). Note that for any hardened gate, the output $outp$ falls after the falling of $outn$, and $outn$ rises after the rising of $outp$. Therefore, the rising (falling) delay of a hardened gate is obtained from the rising (falling) delay of the $outn$ ($outp$) node. After the characterization of all hardened gates, the modified static timing analysis tool was used to compute the delay of the radiation hardened circuits using these two-dimensional delay lookup tables. T_{su} and T_{cq} were obtained using an unhardened D flip-flop for the regular design, and a radiation tolerant flip-flop [2] for the hardened design. Table 7.2 also reports the critical charge value for the radiation hardened design. Column 1 reports the circuit under consideration. Columns 2 and 3 report the clock period (in ps) for a regular area mapped design and the hardened area mapped design. Column 4 reports the percentage delay overhead (or clock period overhead) for the radiation-hardened design. Columns 5 report the critical charge (in fC) for the hardened design, computed as described in Sect. 7.3.3. Note that the Q_{cri} value is obtained for $\tau_\alpha = 150\,\mathrm{ps}$ and $\tau_\beta = 38\,\mathrm{ps}$, as reported in [6]. The smallest gate in *LIB* was used to find this value. Columns 6–9 report the same results as Columns 2–5, but for delay mapped designs. As reported in Table 7.2, the average delay overhead of the proposed radiation hardening approach is 28.9% and 28% for area and delay mapped designs, respectively. Also, the critical charge for the radiation hardened design is a very large value. Traditional radiation hardening approaches such as [6, 10] protect against radiation strikes of at most ~150 fC. For all but one hardened design obtained using the proposed approach, the critical charge is greater 650 fC[4] for $\tau_\alpha = 150\,\mathrm{ps}$ and $\tau_\beta = 38\,\mathrm{ps}$. Therefore, for all practical purposes, the proposed radiation hardening approach provides 90% coverage (i.e., a reduction of the soft error rate by an order magnitude) against very high energy radiation particle strikes.

Table 7.3 reports the area and delay overheads of the radiation hardened designs obtained using the approach proposed in this chapter, to achieve *100% coverage* for both area and delay mapped designs. The overheads reported in Table 7.3 are the percentage area and delay overheads of the hardened circuits, compared with their regular counterparts. Note that the actual area and delay numbers of the regular circuits are reported in Tables 7.1 and 7.2. Table 7.3 shows that for 100% coverage, the proposed hardening approach results in a 97.74% area overhead and a 44.32% delay overhead on average, for area mapped designs. For delay mapped designs, the area overhead is 96.49% and the delay overhead is 47.61%. Note that the area and the delay overheads for 100% radiation tolerance are approximately 50–60% higher than the area and the delay overheads for 90% coverage. Note that this design point is appealing since it protects 100% of the circuit against significantly larger Q_{cri} values

Table 7.3 Area and delay overheads of the proposed radiation hardened design approach for 100% coverage

	Area map		Delay map	
Ckt	% Area ovh.	% Delay ovh.	% Area ovh.	% Delay ovh.
alu2	97.86	39.76	97.21	48.27
alu4	97.87	39.22	97.18	44.42
apex7	97.47	40.56	96.04	41.40
C1355	97.27	45.93	96.37	44.97
C1908	97.41	47.33	95.82	49.13
C3540	98.55	46.83	97.35	47.95
C432	98.19	44.62	97.06	47.32
C499	97.09	45.96	95.91	45.07
C880	97.7	44.90	96.66	46.83
dalu	97.98	49.53	96.4	52.05
frg2	97.4	42.85	95.4	56.31
AVG	97.74	44.32	96.49	47.61

[4] The pulse width of the voltage glitch induced by a radiation particle strike with $Q > 650\,\mathrm{fC}$ saturates to a value of 660 ps.

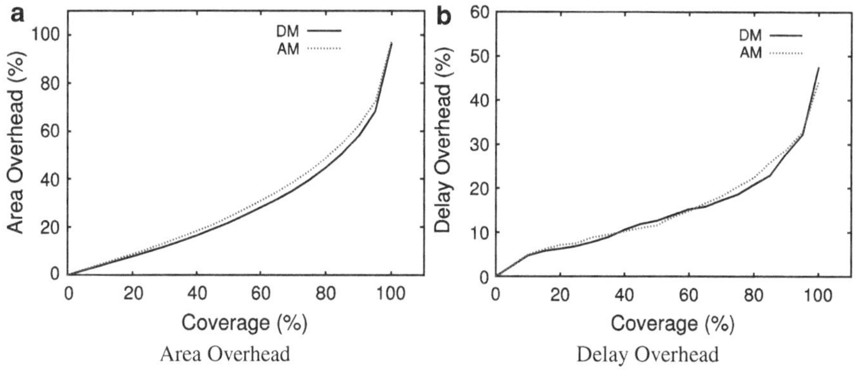

Fig. 7.9 Area and delay overhead of our radiation hardening design approach for different coverage

than has been reported in the literature. Of course, for soft error rate reductions lower than 100%, these overheads can be significantly reduced, described next.

Figure 7.9 shows the average area and delay overheads of the radiation hardened designs, obtained by using the split-output-based hardening approach for different coverage values. In Fig. 7.9, AM (DM) corresponds to area (delay) mapped designs. As shown in Fig. 7.9, initially both the area and delay overheads increase linearly with an increasing coverage value, for coverage less than ~80%. However, for coverage values greater than 80%, the area and delay overheads increase super-linearly with increasing coverage. Thus, the optimal coverage value for the split-output-based hardening approach is about 80%. For 80% coverage, the average area and delay overheads for area (delay) mapped designs are 48.7% and 22.4% (44.7% and 20.8%), respectively.

From Figs. 7.2 and 7.6, it can be concluded that the proposed radiation tolerant standard cells can tolerate high energy radiation particle strikes without affecting the state of gates in their fanout. Also Tables 7.1 and 7.2 show that the circuit radiation hardening technique proposed in this chapter provides good soft error rate reduction (by an order of magnitude) with a modest area overhead of 60% and delay overhead of 29% on average. The critical charge of the hardened circuit obtained using the proposed approach is also a very large value (>650 fC in all but one example), which ensures correct circuit functionality in a heavily radiation prone environment.

7.5 Chapter Summary

This chapter presents a new radiation tolerant CMOS standard cell library, and demonstrates its effectiveness in implementing digital circuits. It is known that if a gate is implemented using only PMOS (NMOS) transistors, then a radiation particle strike can result only in a logic 0–1 (1–0) glitch. This concept was applied to derive radiation hardened standard cells. The radiation hardened gates exhibit an extremely high degree of radiation tolerance and significantly reduced leakage compared with competing approaches. This is validated through circuit simulations at the circuit level. The work presented in this chapter also implemented circuit level hardening using logical masking, to selectively harden those gates in a circuit which contribute most to the soft error failure rate of the circuit. The gates with a low probability of logical masking are replaced by radiation tolerant gates from the new library, such that the digital design achieves 90% soft error rate reduction. Experimental results validate the claims of high radiation tolerance. A 90% reduction in SER is achieved with an area (delay) penalty of 62% (29%) for area mapped designs.

References

1. S. Whitaker, J. Canaris, and K. Liu, "SEU hardened memory cells for a CCSDIS Reed Solomon encoder," *IEEE Transactions on Nuclear Science*, vol. 38, no. 6, pp. 1471–1477, 1991.
2. M. N. Liu and S. Whitaker, "Low power SEU immune CMOS memory circuits," *IEEE Transactions on Nuclear Science*, vol. 36, no. 6, pp. 1679–1684, 1992.
3. J. Canaris, "An SEU immune logic family," in *Proc. of the NASA Symposium on VLSI Design*, Oct 1991, pp. 2.3.1–2.3.11.
4. E. J. Nowak, "Maintaining the benefits of CMOS scaling when scaling bogs down," *IBM Journal of Research and Development*, vol. 46, no. 2-3, pp. 169–186, 2002.
5. Nanoscale integration and modeling (NIMO) group (2007), *ASU Predictive Technology Model [On-line]*, Available: http://www.eas.asu.edu/~ptm.
6. Q. Zhou and K. Mohanram, "Gate sizing to radiation harden combinational logic," *IEEE Transactions on Computer-Aided Design of Integrated Circuits and Systems*, vol. 25, no. 1, pp. 155–166, Jan. 2006.
7. ECSS: European cooperation for space standardization, *Energetic particle radiation*, May 2008, Available: http://www.spenvis.oma.be/spenvis/ecss/ecss09/ecss09.html.
8. J. Feynman, G. Spitale, J. Wang, and S. Gabriel, "Interplanetary proton fluence model: Jpl 1991," *J. Geophys. Res.*, vol. 98, pp. 13281–13294, 1993.
9. Cadence Design Systems, Inc., San Jose, CA, *Envisia Silicon Ensemble Place-and-route Reference Manuals*, Nov 1999.
10. C. Nagpal, R. Garg, and S. P. Khatri, "A delay-efficient radiation-hard digital design approach using CWSP elements," in *Proc. of the Conf. on Design Automation and Test in Europe*, March 2008, pp. 354–359.

Part II
Process Variations

Chapter 8
Sensitizable Statistical Timing Analysis

8.1 Introduction

As described in Chap. 1, with the continuous scaling of the minimum feature sizes of VLSI fabrication processes, variations in key MOSFET and interconnect parameters are increasing at an alarming rate [1, 2, 3]. The increasing variability of device and interconnect parameters makes the task of designing reliable VLSI systems difficult. Thus, it is essential to use a statistical analysis of timing to evaluate the performance of a combinational circuit under variations. Also, design methodologies to implement variation tolerant circuits need to be developed and validated by statistical timing analysis tools.

In this monograph, a sensitizable statistical timing analysis approach is developed, to improve the accuracy of timing analysis under process variations. The sensitizable statistical timing analysis approach is referred to as StatSense, and it is described in this chapter. Two design approaches are also developed to improve the process variation tolerance of combinational circuits and voltage level shifters (which are used in circuits with multiple interacting supply domains), respectively. The process variation tolerant design approach for combinational circuits is discussed in the next chapter. In Chap. 10, the process variation tolerant voltage level shifter design is described.

In recent times, statistical static timing analysis (SSTA) has received significant attention in both academe and industry. Although a lot of research has suggested that SSTA is essential for timing closure in contemporary VLSI design, this new method of timing analysis has not been readily accepted by chip designers. It is not just the reticence of designers toward adopting a new design methodology that is preventing/slowing the adoption of this new timing approach. There is also a legitimate concern that the results of SSTA tend to be overly pessimistic. Besides, SSTA takes longer to run. Also, the adoption of SSTA requires a greater effort during the gate library characterization of VLSI design phase. Designers are hence skeptical about the benefits of this new timing analysis methodology.

The are many factors that contribute to the inaccuracies of SSTA, and many of them are dependent on the method used for the analysis. Some of these factors are:

1. Spatial correlations of process variations

R. Garg and S. P. Khatri, *Analysis and Design of Resilient VLSI Circuits: Mitigating Soft Errors and Process Variations*, DOI 10.1007/978-1-4419-0931-2_8,

2. Correlations of the delays of different circuit paths
3. In block-based SSTA [4], the PDF (probability density function) that results from the *maximum* operation of two PDFs is approximated to have a Gaussian distribution
4. False paths in the circuit
5. Representation of gate delay distribution. Usually, a single distribution for each pin to output delay for a gate is used. However, accuracy would be improved if one distribution is used for each input vector transition

The last two issues mentioned above are also sources of pessimism for current SSTA tools. Most static timing analysis (STA) tools (and their statistical counterparts) do not consider false paths. Also these tools assume that the delay distributions of all gates in a design are Gaussian. However, the delay distribution of a gate is not necessarily Gaussian. In fact, the delay distribution for a multi-input gate is Normal for each input vector transition that causes a change on the gate output.

The StatSense approach proposed in this chapter deals with these two issues. StatSense consists of two phases. In the first phase, a set of N logically sensitizable vector transitions that result in the largest delays for a circuit are obtained. In the second phase, these delay-critical sensitizable input vector transitions are propagated using a Monte-Carlo-based technique, to obtain a delay distribution at the outputs. The specific input transitions at any gate are known after the first phase, and so the gate delay distributions corresponding to these input transitions are utilized in the second phase. The second phase performs Monte-Carlo-based SSTA, using the appropriate gate delay distribution corresponding to the input transition for each gate. In this way, StatSense also implicitly considers path correlations and does not approximate gate delay PDFs to Gaussian distributions (since it does not require the computation of the MAX of two PDFs).

The rest of the chapter is organized as follows. Section 8.2 briefly discusses previous work on statistical timing analysis of combinational circuits. The proposed StatSense approach is described in Sect. 8.3. Experimental results are presented in Sect. 8.4, followed by a chapter summary in Sect. 8.5.

8.2 Related Previous Work

The idea of statistical timing analysis has been a subject of research for several years. Some of the early works in this field include [5, 6]. The recent growth of interest in this field has been driven primarily due to the fact that process variations are growing larger and less systematic.

Most of the techniques that perform statistical timing analysis are based on the principles of STA. STA computes the (pessimistic) worst case delay of a circuit by propagating the worst case arrival time at the nodes of a circuit (in a topological manner from inputs to outputs). The arrival time at the output of each gate is the MAX of the SUM of the gate's delay and the arrival time of its inputs. Since most STA approaches utilize STA to propagate delay, they are often called SSTA approaches.

There are two broad classes of SSTA algorithms – path-based and block-based. In path-based algorithms, a set of paths is selected for a detailed statistical analysis, which is performed by Monte-Carlo techniques. In each iteration, delay computation is performed in a breadth first manner from circuit inputs to outputs, using STA. In block-based algorithms, delay distributions are propagated by traversing the circuit under consideration in a levelized breadth-first manner. The fundamental operations in a block-based SSTA tool are the SUM and the MAX operations. Most block-based SSTA algorithms rely on efficient ways to implement these SUM and MAX operations for delay distributions, rather than for discrete delay values (which STA uses). Block-based algorithms tend to be faster, while path-based algorithms are more accurate and provide more realistic statistical timing estimates [7].

In [4], the authors present a technique to propagate probability density functions (PDFs) through a circuit in the same manner as arrival times of signals are propagated during STA. Using principal component analysis (PCA), they also demonstrate the ability to handle spatial correlations of process parameters. While the SUM operation used (for adding 2 Gaussian distributions) yields another Gaussian distribution, the MAX of 2 or more Gaussian distributions is not a Gaussian distribution in general. For the sake of simplicity and ease of calculation, the authors of [4] approximate the MAX of 2 or more Gaussian distributions to be Gaussian as well.

In [7], a canonical first-order delay model is proposed, and an incremental block-based timing analyzer is used to propagate arrival times and required times through a timing graph in this canonical form. One of the major contributions of the algorithm proposed in [7] is that it allows the statistical timing engine to be used in an incremental manner.

In [8,9,10], the authors note that accurate SSTA can become exponential. Hence, they propose faster algorithms that compute bounds on the exact result.

In [11], a block-based SSTA algorithm is discussed. By representing the arrival times as cumulative distribution functions (CDFs) and the gate delays as PDFs, the authors claim to have an efficient method to do the SUM and MAX operations. They decompose the CDF into a sum of ramps and the PDF into a sum of step signals. This discretization helps to make the SUM and MAX operation more efficient. The accuracy of the algorithm can be adjusted by choosing more discretization levels. Reconvergent fanouts are handled through a statistical subtraction of the common mode.

In [12], the authors propagate gate delay distributions (PDFs) through a circuit. The key contribution of [12] is that PDFs are discretized to help make the operation more efficient. Here too, the accuracy of the result is dependent on the discretization.

The common theme in all of the above works is that they are based on the STA framework. Hence, only the structurally long paths are identified through these algorithms. The authors of [13] identify this deficiency and come up with a SSTA flow that considers false paths. Their flow consists of two phases. In the first phase, a regular SSTA (using worst-case statistical timing information) is performed to identify the structurally long paths. Each of these paths are then checked to see if they are logically sensitizable. In the second phase of the flow, another SSTA is performed for just the logically sensitizable paths, to check if they are "timingly true." A path is timingly true if the transitions on the path cannot be invalidated by other off-path inputs. The SSTA is done using Monte-Carlo-based techniques. Although the authors of [13] reduce pessimism by considering false paths, they do not address the pessimism that arises from using a single Gaussian gate delay distribution for any gate.

In [14], a bound-based technique was proposed to identify the top timing-violating paths in a circuit under variability. The authors of [14] verified the correctness and accuracy of their approach and found that their approach finds delay critical paths of a circuit under variations with good fidelity. Note that the approach of [14] can be extended to obtain the sensitizable delay critical input vector transitions, which are required in the second phase of the StatSense approach.

Thus, traditional approaches for STA of a circuit tend to be overly pessimistic, and hence pose tighter design constraints on the VLSI circuit designers. Therefore, more accurate STA tools are desired, to simplify the designers task.

8.3 Proposed Sensitizable Statistical Timing Analysis Approach

The StatSense approach eliminates false paths and also accounts for the fact that the delay of a gate has different normal distributions for different input transitions, which cause an output transition. As mentioned earlier, the StatSense approach consists of two phases. In the first phase, a set of N logically sensitizable vector transitions that result in the largest delays for the circuit are obtained. In the second phase, these delay-critical sensitizable input vector transitions are propagated using a Monte-Carlo-based technique, to obtain a delay distribution at the outputs. The specific input transitions at any gate are known after the first phase, and so the gate delay distributions corresponding to these input transitions are utilized in the second phase. The second phase performs Monte-Carlo-based SSTA, using the appropriate gate delay distribution corresponding to the input transition for each gate.

In the remainder of this section, these two phases are described, along with a discussion on how input arrival times are propagated for any gate.

8.3.1 Phase 1: Finding Sensitizable Delay-critical Vector Transitions

To ensure that the time is not needlessly spent on performing statistical analysis on false paths, a user-specified number N of sensitizable vector transitions (that result in the largest delays for the circuit) are first obtained. This is done using the *sense* [15] package in SIS [16]. *Sense* uses a boolean satisfiability (SAT) [17,18,19] solver to verify if a particular delay (initially set to the delay found from a STA run) is sensitizable. As a consequence, *sense* is NP complete. Efficient implementations of this algorithm [15,20] exist, and have been demonstrated to work on large designs. If there is no satisfiable input vector that produces the delay obtained by the STA, the delay value is reduced in steps until a delay D value is reached that has a satisfying vector (a vector on the primary inputs that has a delay D) is obtained. In its original implementation, *sense* returns only the maximum sensitizable delay of the circuit. On the basis of the theory of the sensitizable timing analysis methodology of *sense* [15], the resulting delay that is reported by the approach is the largest delay for the design to reach a stable output state. The *sense* routine was modified to return the final vector on the primary inputs, as well as all the possible initial vectors on the primary inputs, that cause the maximum sensitizable delay. A change from any initial vector to a final vector is referred to as a *vector transition*. The set of N input transitions is stored in an array for use in the second phase of the proposed statistical timing flow.

After obtaining the first largest sensitizable delay vector, the complement of this delay vector is inserted (as a SAT clause) in *sense's* SAT routine, and then *sense* is run again to get the next critical vector. For example, if the vector V that produced the largest sensitizable delay D at output z is $a\overline{b}c$ (where a, b, and c are the primary inputs involved in the sensitizable critical delay transition), then the clause added is $(\overline{a}+b+\overline{c}+\overline{z})$. When *sense* is called again, it returns the vector transition resulting in the next largest sensitizable delay. Note that different vector transitions sensitize different structural paths. This process is continued until the top N delay-critical, sensitizable vector transitions are collected. Note that the value of N is specified by the user. It can be decided based on the desired accuracy, and the time available for computation.

In [13], the authors find a set of logically sensitizable paths from a given set of structurally long paths (found from an initial STA). Then from this set, they separately find a subset of paths that are "timingly true."

In the proposed approach, *sense* returns vector transitions that are logically sensitizable *and* timing true. There is no separation of the two properties (logic sensitizability and timing trueness). *Sense* performs its analysis using nominal gate delay values. The statistical analysis is done only on the N primary input vector transitions that *sense* declared to be critical and sensitizable.

Once *sense* returns a set of N critical vector transitions, these vector transitions are then used to find the statistical distribution of the circuit delay due to these vector transitions. A primary input vector transition may induce an input transition on each gate of the design, for which the appropriate gate delay distribution is selected for further statistical processing. Also, the arrival times for the gate are propagated in a manner that exploits the fact that the input transition at the gate are known. This is explained in the following section.

8.3.2 Propagating Arrival Times

In a regular STA, the structurally worst delay is obtained. However, StatSense uses the fact that the specific transitions at the inputs of a gate, which cause the output node to switch, are known. The details of how this is done is explained with the

Table 8.1 Transitions for a NAND gate that cause its output to switch

Rising transition #	ab → ab	Delay(ps)
1	11 → 00	30.5
2	11 → 01	50.5
3	11 → 10	53.0
Falling transition #	ab → ab	Delay(ps)
1	00 → 11	55.3
2	01 → 11	46.5
3	10 → 11	42.7

example of a NAND2 gate, which is assumed to be part of a larger circuit. First, consider just the nominal delay of a NAND2 gate.

Table 8.1 is a list of input transitions that cause the output of the NAND gate to change its logic value. Let $\mathrm{AT}_i^{\mathrm{fall}}$ denote the arrival time of a falling signal at node i and $\mathrm{AT}_i^{\mathrm{rise}}$ denote the arrival time of a rising signal at node i.

In the case of regular STA, the rising time (delay) at the output c of a NAND2 gate is calculated as

$$\begin{aligned}\mathrm{AT}_c^{\mathrm{rise}} = \mathrm{MAX}[(&\mathrm{AT}_a^{\mathrm{fall}} + \mathrm{MAX}(D_{11\to 00}, D_{11\to 01})),\\ &(\mathrm{AT}_b^{\mathrm{fall}} + \mathrm{MAX}(D_{11\to 00}, D_{11\to 10}))],\end{aligned}$$

where $D_{xy\to pq}$ is the delay of the output when the inputs change from xy to pq. Also, $\mathrm{MAX}(D_{11\to 00}, D_{11\to 01})$ is often referred to as the pin-to-output rising delay from the input a, while $\mathrm{MAX}(D_{11\to 00}, D_{11\to 10})$ is referred to as the pin-to-output rising delay from the input b.

Similarly, in STA, the falling time (delay) at the output c of a NAND2 gate is given by

$$\begin{aligned}\mathrm{AT}_c^{\mathrm{fall}} = \mathrm{MAX}[(&\mathrm{AT}_a^{\mathrm{rise}} + \mathrm{MAX}(D_{00\to 11}, D_{01\to 11})),\\ &(\mathrm{AT}_b^{\mathrm{rise}} + \mathrm{MAX}(D_{00\to 11}, D_{10\to 11}))],\end{aligned}$$

where $\mathrm{MAX}(D_{00\to 11}, D_{01\to 11})$ is often referred to as the pin-to-output falling delay from the input a, while $\mathrm{MAX}(D_{00\to 11}, D_{10\to 11})$ is referred to as the pin-to-output falling delay from input b.

For example, if the worst case falling or rising arrival time at inputs a and b was 10 ps and 35 ps, respectively, then the rise delay at c would be calculated to be = $\mathrm{MAX}(10 + 50.5, 35 + 53.0) = 88.0$ ps as shown in Fig. 8.1. Similarly for a falling c output, the delay would be $\mathrm{MAX}(10 + 55.3, 35 + 55.3) = 90.3$ ps (shown in Fig. 8.1). However, this is a pessimistic method of calculating the delay. The StatSense approach attempts to remove some of this pessimism.

First consider the rising output. The output of the NAND2 gate switches high when any of the two inputs switches low. From the output of *sense*, the actual vector transition that causes the largest delay for a given circuit can be found. This primary

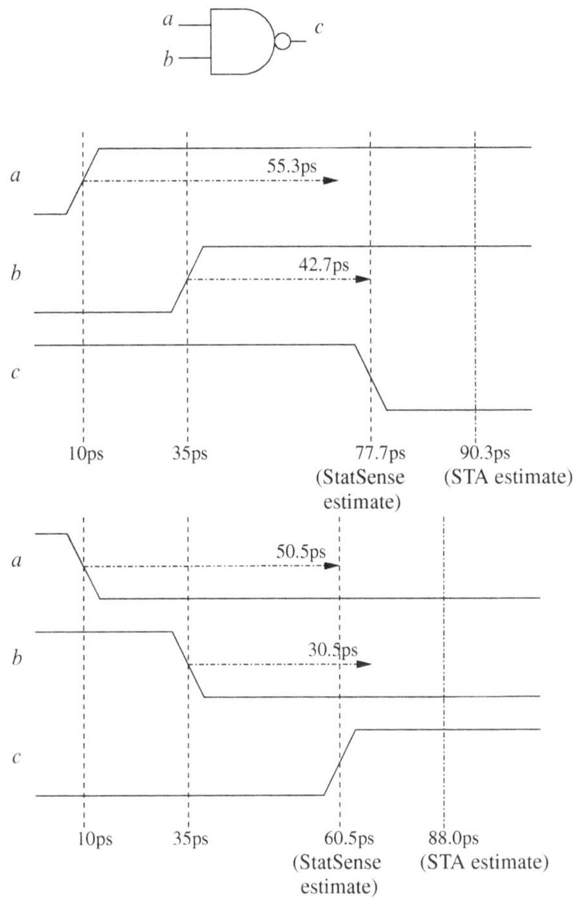

Fig. 8.1 Arrival time propagation using a NAND2 gate

input vector transition induces a transition on the gate inputs. Assume that this input transition was 11 → 00 for the NAND2 gate. A naive way of calculating the delay would be to state that the delay would be given by

$$AT_c^{rise} = MAX(AT_a^{fall}, AT_b^{fall}) + D_{11\to 00}.$$

Assuming again that the arrival times at inputs a and b were 10 ps and 35 ps, respectively, the delay would be then be calculated as $MAX(10, 35) + 30.5 = 65.5$. However, it is known that the output would start switching before 65.5, since signal a arrives earlier than signal b. As a result, the gate effectively goes through the transition 11 → 01 → 00 rather than 11 → 00 directly. Note that the output of the NAND2 gate falls for the vector 01 as well. Hence, StatSense calculate the delay to be

$$AT_c^{rise} = MIN((AT_a^{fall} + D_{11\to 01}), (AT_b^{fall} + D_{11\to 00})).$$

In this example, the delay estimated by StatSense is hence $\text{MIN}(10 + 50.5, 35 + 30.5) = 60.5$ (shown in Fig. 8.1). Note that the minimum of two delays is used in this case since any one input falling causes the output to switch. Also note that the delay calculated (60.5 ps) is much smaller than the worst case delay calculated using regular STA (88.0 ps) (shown in Fig. 8.1). The reduction in pessimism in the proposed approach occurs due to the fact that the information about the input transition for the gate is available.

Now consider the case of the falling output. The output of the NAND2 gate switches low only when both the inputs switch high. Again, StatSense uses the fact that *sense* provides the actual vector transition that caused the critical delay. Assume that the induced input transition for the NAND2 gate was $00 \rightarrow 11$. A naive way of calculating the delay would be to state that the delay is

$$\text{AT}_c^{\text{fall}} = \text{MAX}(\text{AT}_a^{\text{rise}}, \text{AT}_b^{\text{rise}}) + D_{00 \rightarrow 11}.$$

Assuming again that the arrival times at inputs a and b were 10 ps and 35 ps, respectively, the delay would be calculated as $\text{MAX}(10, 35) + 55.3 = 90.3$. However, a arrives earlier than b. As a result, the gate effectively goes through the transition $00 \rightarrow 10 \rightarrow 11$ rather than $00 \rightarrow 11$ directly. Hence, in the StatSense approach, the delay is calculated to be

$$\text{AT}_c^{\text{fall}} = \text{MAX}((\text{AT}_a^{\text{rise}} + D_{00 \rightarrow 11}), (\text{AT}_b^{\text{rise}} + D_{10 \rightarrow 11})).$$

In this example, the delay is hence $\text{MAX}(10 + 55.3, 35 + 42.7) = 77.7$. Note that the maximum of two delays is used in this case since both inputs need to switch to cause the output to switch. Also note that the delay calculated (77.7 ps) is smaller than the worst case delay calculated using regular STA (90.3 ps) as shown in Fig. 8.1.

These results are shown graphically in the Figs. 8.2 and 8.3. These plots show the arrival time of the output c of a NAND2 gate, for the $00 \rightarrow 11$ and $11 \rightarrow 00$ transitions, respectively. The arrival time of one of the inputs a is fixed to zero, and the arrival time of the other input b swept between -100 ps and 100 ps. The propagated delays are shown for STA and StatSense, along with the delay found by SPICE [21]. Figures 8.4 and 8.5 show the same data for the NOR2 gate. As can be seen from these plots, the method that is used by StatSense to calculate the arrival times for multiple switching inputs matches SPICE quite accurately and is significantly better (less pessimistic) than a traditional STA method for computing arrival times.

Similarly, equations are derived to calculate the arrival times for an arbitrary gate, depending on the input transitions at that gate. Consider a NAND3 gate with inputs $\{a, b, c\}$. First, assume that the inputs of the NAND3 gate change as follows:

$$000 \rightarrow 100 \rightarrow 110 \rightarrow 111$$

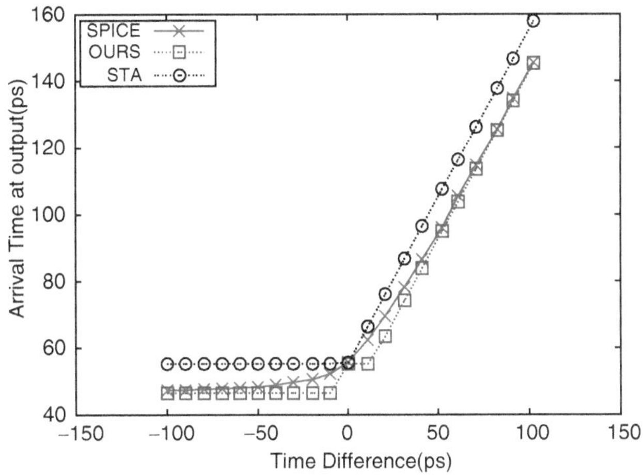

Fig. 8.2 Plot of arrival times at output of NAND2 gate calculated through various means for the transition 00 → 11

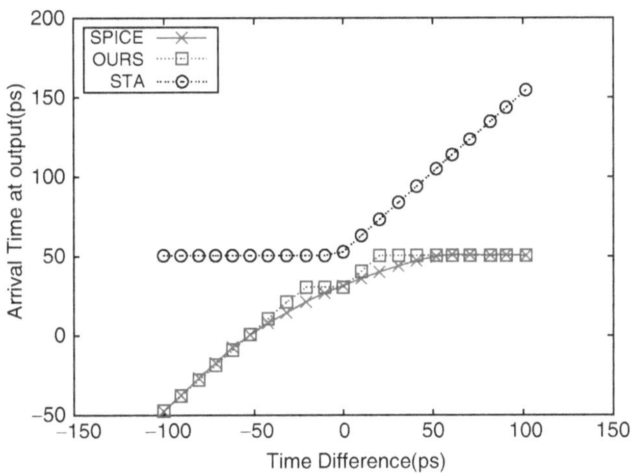

Fig. 8.3 Plot of arrival times at output of NAND2 gate calculated through various means for the transition 11 → 00

The output of the NAND3 gate switches low only when the inputs are 111. Hence the delay of the gate would be calculated as follows:

$$\mathrm{AT}_{\mathrm{out}}^{\mathrm{fall}} = \mathrm{MAX}[(\mathrm{AT}_a^{\mathrm{rise}} + D_{000\to 111}), \tag{8.1}$$

$$(\mathrm{AT}_b^{\mathrm{rise}} + D_{100\to 111}), \tag{8.2}$$

$$(\mathrm{AT}_c^{\mathrm{rise}} + D_{110\to 111})].$$

Now consider a NAND3 gate with its output rising. Let the inputs change as below

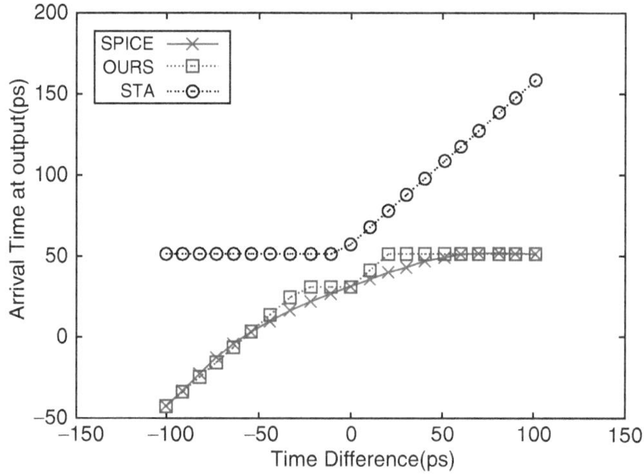

Fig. 8.4 Plot of arrival times at output of NOR2 gate calculated through various means for the transition 00 → 11

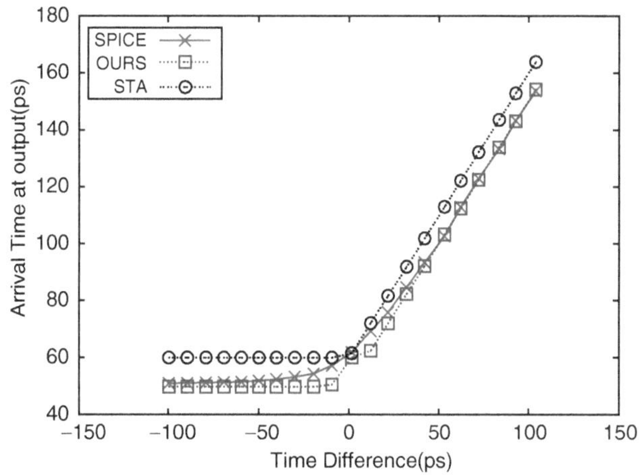

Fig. 8.5 Plot of arrival times at output of NOR2 gate calculated through various means for the transition 11 → 00

$$111 \rightarrow 011 \rightarrow 001 \rightarrow 000$$

In this case, the output of the NAND3 gate starts switching high when at least one of the inputs is logic 0. Hence, the delay of the gate would be calculated as:

$$\begin{aligned} \mathrm{AT}_{\mathrm{out}}^{\mathrm{rise}} = \mathrm{MIN}[(\mathrm{AT}_a^{\mathrm{fall}} &+ D_{111\rightarrow 011}), && (8.3)\\ (\mathrm{AT}_b^{\mathrm{fall}} &+ D_{111\rightarrow 001}), && (8.4)\\ (\mathrm{AT}_c^{\mathrm{fall}} &+ D_{111\rightarrow 000})]. \end{aligned}$$

An extension to handling delay distributions is easily done by simply considering the distribution to be made of several distinct delay values, obtained from the PDF of the gate delay.

8.3.3 *Phase 2: Computing the Output Delay Distribution*

In the second phase of StatSense, Monte-Carlo analysis is performed on the sensitizable vector transitions that result in the largest delays for the circuit (which were computed in the first phase, described in Sect. 8.3.1). In each of the STA runs for Monte-Carlo analysis, arrival times are propagated as described in Sect. 8.3.2. Since the primary input vector transitions may induce transitions on the input of each gate, the delay distribution of the gate for the corresponding gate input transition is used. A random value of the gate delay is computed from this distribution. This is done for each gate in the circuit. Finally, STA is performed using these delay values. The resulting maximum delay over all the outputs is used to compute the worst case delay distribution of the circuit.

In a NAND2 gate, there are three different input rising transitions that cause an output falling transition (these are shown in the bottom half of Table 8.1). For any iteration of STA, if the value of delay for one of the three transitions (say 00 $\rightarrow$ 11) is chosen to be $\mu_{00\rightarrow 11} + n\sigma_{00\rightarrow 11}$, then the value of the other two transitions (01 $\rightarrow$ 11, 10 $\rightarrow$ 11) used is $\mu_{01\rightarrow 11} + n\sigma_{01\rightarrow 11}$ and $\mu_{10\rightarrow 11} + n\sigma_{10\rightarrow 11}$, respectively.

8.4 Experimental Results

To demonstrate the effectiveness of the StatSense approach, it was tested for several circuits from the ISCAS89 and MCNC91 benchmark suite. A 0.1 μm BPTM process [22] model card was used for all SPICE [21] simulations. A standard cell library *LIB* was designed, which consisted of 8 cells. The 8 cells were INV2X, INV4X, NAND2, NAND3, NAND4, NOR3, NOR3, NOR4.

All standard cells in *LIB* were precharacterized to construct a table of values for the mean and standard deviation of the delay of *each transition* (that causes a change in the output). This precharacterization was done for a set of load capacitance values. This precharacterization was done using SPICE. The parameters considered to be varying, along with their variations, are given in Table 8.2. As reported in this table, all parameters were modeled such that their σ is 5% of their μ. The threshold voltages and the channel lengths of the devices in a gate were assumed to vary in the same manner. Thus, all process parameters within a gate were assumed to be perfectly correlated.

The characterization results for a NAND2 gate (with a load capacitance of 6 fF) are shown in Figs. 8.6 and 8.7. Figure 8.6 shows the delay histogram for the three vector transitions, which result in a rising output. These vector transitions are

Table 8.2 Parameters with their variation

Parameter	Nominal value	σ
L	0.1 μ	0.005 μ
V_{TN}	0.2607 V	0.013 V
V_{TP}	0.3030 V	0.01515 V

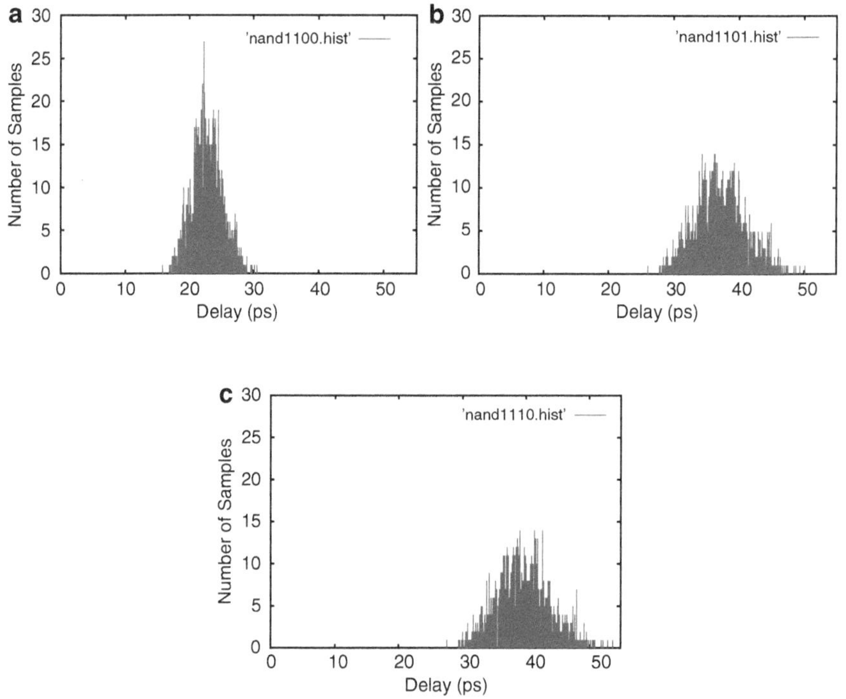

Fig. 8.6 Characterization of NAND2 delay for all input transitions which cause a rising output. (**a**) 11 → 00, (**b**) 11 → 01, and (**c**) 11 → 10

11 → 00, 11 → 01, and 11 → 10. Note that each of these vector transitions exhibit different output delay distributions. Similarly, Fig. 8.7 shows the delay histogram for the three vector transitions, which result in a falling output. These vector transitions are 00 → 11, 01 → 11, and 10 → 11. Note that each of these vector transitions also exhibit different output delay distributions. The mean and standard deviation of the distributions of each of these vector transitions were computed and used in the second phase of the proposed algorithm.

During the timing analysis phase of the StatSense approach, the mean and standard deviation of the delay for a given load capacitance value was obtained by interpolating between the capacitance values for which the precharacterization was performed.

Next, the first phase of the StatSense approach was carried out. *Sense* was used to find the top few sensitizable critical delays and their corresponding input vector

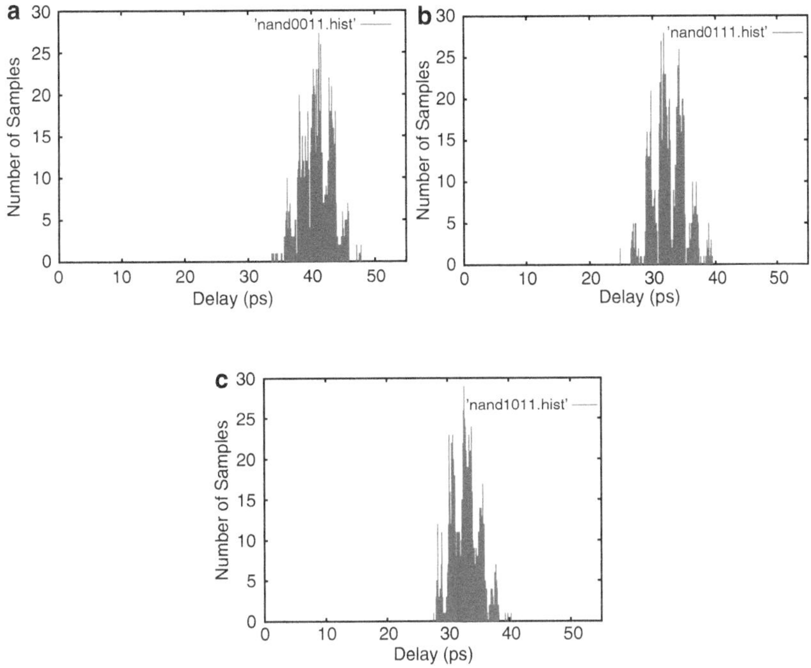

Fig. 8.7 Characterization of NAND2 delay for all input transitions which cause a falling output. (**a**) 00 → 11, (**b**) 01 → 11, and (**c**) 10 → 11

transitions. The result of the first phase of the StatSense approach is a set of vector transitions on the primary inputs of the circuit. The experiments were performed for $N = 75$ (or 50 or 25) primary input vector transitions that result in the largest circuit delay.

For the second phase of the StatSense approach, these transitions were propagated throughout the circuit. Since the input transition at each gate is known, the arrival time propagation methodology explained in Sect. 8.3.2 was used to compute the arrival time at the gate output. The output delay of the circuit was obtained by performing a linear traversal of the circuit in levelized order. This step of propagating circuit delays is done 1,000 times (or as many times as is required to get a reasonably stable and accurate estimate of the mean and standard deviation of the maximum delay of the circuit). For each of these 1,000 iterations, a random value of delay is chosen for each gate, for the relevant input vector transitions for that gate. This random value is chosen from a Gaussian distribution with a μ and σ derived from the precharacterized table of values for each gate, for the appropriate input transition at that gate. Note that the μ and σ used for any gate correspond to the vector transitions that appear at that gate when the primary input vector transition is applied.

Tables 8.3 and 8.4 describe the results of experiments conducted to compare StatSense with SSTA. The major goal of the work presented in this chapter is to make SSTA more accurate. Hence, the StatSense approach was compared with

Table 8.3 Comparison of SSTA and StatSense for 75 input vector transitions

	SSTA				StatSense 75					
Ckt	μ (ps)	σ (ps)	$\mu + 3\sigma$	Time	μ (ps)	σ (ps)	$\mu + 3\sigma$	Ratio	Time	Ratio
alu2	1,008.39	19.08	1,065.63	278.8	701.30	7.46	723.68	0.68	5,095.41	17.78
alu4	1,234.77	18.21	1,289.4	560.2	803.56	7.45	825.91	0.64	6,607.43	11.8
apex5	539.04	19.29	596.91	727.6	464.68	5.84	482.2	0.81	1,481.91	2.04
apex6	680.51	10.95	713.36	632.2	538.27	5.41	554.5	0.78	1,673.73	2.64
apex7	489.85	8.3	514.75	180.66	490.20	5.24	505.92	0.98	323.9	1.79
C499	737.00	11.29	770.87	419.4	652.45	5.68	669.49	0.87	636.46	1.52
C880	1,037.51	13.56	1,078.19	324.9	909.20	7.40	931.4	0.86	651.8	2.01
C1355	714.82	8.59	740.59	484.8	445.46	4.49	458.93	0.62	673.81	1.39
cordic	669.99	8.60	695.79	657.0	620.17	7.14	641.59	0.92	1,093.63	1.6
i6	496.16	22.80	564.56	353.0	495.55	7.89	519.22	0.92	836.8	2.36
i7	496.25	21.76	561.53	514.3	499.26	7.88	522.9	0.93	1,019.4	1.98
i10	1,858.92	18.72	1,915.08	1,900.12	1,323.72	10.25	1,354.47	0.71	3,584.5	1.89
rot	781.23	13.75	822.48	571.0	542.55	6.71	562.68	0.68	1,317.9	2.31
x1	319.34	10.40	350.54	261.5	303.41	6.14	321.83	0.92	321.68	1.23
AVG								0.81		2.49

Table 8.4 Comparison of SSTA and StatSense for 50 and 25 input vector transitions

Ckt	StatSense 50						StatSense 25					
	μ (ps)	σ (ps)	$\mu + 3\sigma$	Ratio	Time	Ratio	μ (ps)	σ (ps)	$\mu + 3\sigma$	Ratio	Time	Ratio
alu2	701.12	7.38	723.26	0.68	4,018.89	14.02	701.25	7.69	724.32	0.68	1,232.5	4.42
alu4	800.62	7.80	824.02	0.64	3,386.8	6.04	799.77	7.93	823.56	0.64	3,217.9	5.74
apex5	463.58	6.07	481.79	0.81	981.2	1.34	460.52	7.46	482.9	0.81	526.7	0.72
apex6	538.48	5.38	554.62	0.78	1,096.42	1.73	536.92	5.98	554.86	0.78	787.7	1.24
apex7	490.10	5.26	505.88	0.98	233.17	1.29	490.05	5.42	506.31	0.98	124.5	0.69
C499	650.35	6.18	668.89	0.87	481.6	1.15	646.23	7.08	667.47	0.87	238.8	0.57
C880	909.13	7.48	931.57	0.86	416.6	1.28	907.65	7.80	931.05	0.86	227.07	0.70
C1355	443.67	4.85	458.22	0.62	578.1	1.2	440.86	5.64	457.78	0.62	291.8	0.60
cordic	613.14	5.69	630.21	0.91	657.23	1.00	612.19	6.03	630.28	0.91	355.3	0.54
i6	493.86	8.39	519.03	0.92	609.5	1.73	488.51	9.70	517.61	0.92	293.79	0.83
i7	496.08	8.72	522.24	0.93	494.9	0.96	489.83	10.06	520.01	0.93	366.6	0.65
i10	1,323.24	10.55	1,354.89	0.71	2,319.0	1.22	1,316.12	11.71	1,351.25	0.71	1,188	0.62
rot	540.15	7.33	562.14	0.68	1,343.6	2.35	535.59	8.41	560.82	0.68	810.4	1.42
x1	303.51	6.10	321.81	0.92	277.14	1.06	303.37	6.28	322.21	0.92	141.6	0.54
AVG				0.81		1.74				0.81		0.95

Monte-Carlo-based SSTA, which is considered to be *most accurate* [23]. It is well known that block-based SSTA sacrifices some accuracy for speed due to approximations when propagating PDFs (especially when computing the MAX of two or more PDFs). Both StatSense and Monte-Carlo-based SSTA were implemented in *SIS* [16]. These data (the mean and the standard deviation of the delay of each transition) obtained from characterization of all standard cells in *LIB* were used for both StatSense and Monte-Carlo-based SSTA. The SSTA experiments in this table were conducted using 10,000 iterations. The StatSense iterations were computed using 1,000 iterations per primary input vector transition. StatSense computes the μ and σ of the delay of a circuit by taking the *statistical maximum* of the delay distributions of all input vector transitions. The statistical maximum was computed as follows. First, a random delay value for each input vector transitions of a circuit is obtained using their corresponding delay distributions. Then the maximum delay value across all vector transitions is selected to obtain the delay of the circuit. This process is repeated a large number of times (10,000) to obtain the final delay distribution of the circuit. Note that in the all results presented in this section, the average of the runtime ratios are computed using a geometric mean, because of the high variability of these ratios. Also note that the runtimes include the time required to obtain the delay-critical vector transitions (Phase 1 of the StatSense approach) and the time required to perform Monte-Carlo iterations for all input vector transitions obtained from Phase 1. In Table 8.3, Column 1 lists the circuit under consideration. Columns 2 through 4 list the μ, σ, and $\mu + 3\sigma$ delays (in ps) returned by SSTA. Column 5 lists the SSTA runtime. All runtimes in this table are in seconds. Columns 6 through 11 list the results for StatSense, when $N = 75$ input vector transitions were simulated. Columns 6 through 8 list the μ, σ, and $\mu + 3\sigma$ delays (in ps) returned by StatSense. Column 9 reports the ratio of the $\mu + 3\sigma$ value returned by StatSense, compared with that returned by SSTA. *Note that StatSense, on average, returns a much lower worst case circuit delay (the $\mu + 3\sigma$ delay) than SSTA. This illustrates the pessimism of SSTA, and validates the claim that StatSense reduces this pessimism.* Columns 10 and 11, respectively, list the runtime for StatSense and the ratio of this runtime vs. the runtime of SSTA. On average, note that StatSense (run with 75 input vector transitions) requires about 2.5× more runtime than SSTA. In Table 8.4, Columns 2 through 7 (and 8 through 13) have the same information as Columns 6 through 11 of Table 8.3, except that the StatSense simulations for these columns were performed using 50 (and 25) input vector transitions (which result in the largest sensitizable circuit delay). The StatSense approach with 50 (25) input vector transitions is referred to as StatSense50 (StatSense25). The purpose of this experiment was to verify if the StatSense runtime can be reduced by simulating fewer input vector transitions. By comparing Columns 9 of Table 8.3 with columns 5 and 11 of Table 8.4, it can be observed that there is no appreciable loss of fidelity when 25 input vector transitions are used instead of 75 or 50. The worst case circuit delay (the $\mu + 3\sigma$ delay), averaged over all designs, is almost identical in all cases. The benefit of using 25 input vector transitions is indicated in Column 13 of Table 8.4, which shows that on average, StatSense (with 25 input vector transitions) requires 5% less runtime than SSTA.

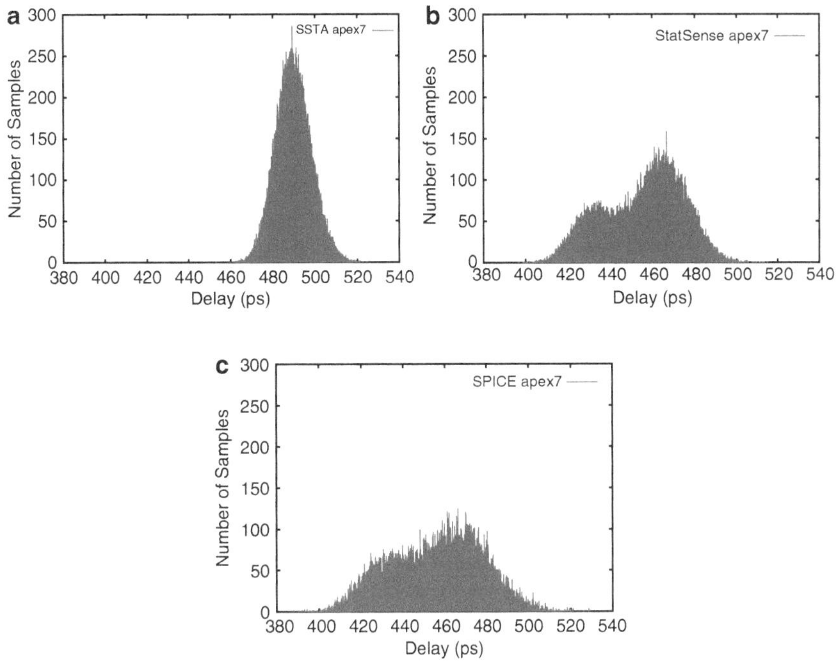

Fig. 8.8 Delay histograms for (**a**) SSTA, (**b**) StatSense, and (**c**) SPICE (for *apex*7)

In spite of the fact that SSTA conducts 10,000 STA iterations, and StatSense conducts 75,000 (or 50,000 or 25,000 for StatSense50 and StatSense25, respectively) iterations, the runtime of StatSense is not 7.5× (or 5× or 2.5×) that of SSTA but rather it is 2.49× (or 1.74× or 0.95×) that of STA. This is because StatSense performs an event driven delay simulation. Whenever there is no transition at the output of a gate g, delay computations for gates in the fanout of g are avoided. This pruning is not possible in SSTA.

Figure 8.8 illustrates the delay histogram obtained by SSTA (with 50,000 STA iterations) along with the delay histogram obtained by StatSense and SPICE (with 50 input vector transitions simulated). These histograms were obtained for the *apex7* example. For each input vector transition in StatSense and SPICE, 1,000 Monte-Carlo iterations of delay computation were performed. This figure shows how the pessimism of SSTA is alleviated by StatSense. This figure also shows that the delay distribution obtained by StatSense closely matches with the delay distribution obtained by SPICE. However, the SSTA method of [13] (which is the best know previous approach) results in 12% higher delay values than SPICE.

As mentioned earlier, StatSense addresses two sources of pessimism in SSTA. These two sources are: false paths in a circuit, and the representation of gate delay distributions. Tables 8.3 and 8.4 report the results obtained when both these issues are addressed simultaneously. To evaluate the accuracy gained by each of these issues separately, Monte-Carlo-based SSTA simulations were performed on 50 sensitizable paths (the same paths which are obtained by StatSense50). In other words, Monte-Carlo-based SSTA analysis was performed on sensitizable paths of the circuits (after eliminating the false paths). Table 8.5 compares the results obtained from Monte-Carlo-based SSTA with and without false path elimination, and StatSense (which eliminates false paths and also models input vector transition based gate delays) with 50 input vector transitions. For each of the 50 input vector transitions, 1,000 Monte-Carlo simulations were performed. In Table 8.5, Column 1 reports the circuit under consideration. Columns 2 through 4 list the μ, σ, and $\mu + 3\sigma$ delays (in ps) returned by SSTA (without false path elimination). Columns 5 through 7 list the μ, σ, and $\mu + 3\sigma$ delays (in ps) returned by StatSense50. Column 8 reports the ratio of the $\mu + 3\sigma$ value returned by StatSense50, compared with that returned by SSTA. Columns 10 through 12 list the μ, σ, and $\mu + 3\sigma$ delays(in ps) returned by performing Monte-Carlo-based SSTA on 50 sensitizable paths, which are obtained by StatSense50 (henceforth referred to as "SSTA with false path elimination"). Column 13 reports the ratio of the $\mu + 3\sigma$ value returned by SSTA without false paths, compared with that returned by SSTA (with false paths). Observe from Table 8.5 that StatSense50 on average reduces the error in the estimation of the worst case circuit delay by 19% compared with the Monte-Carlo-based SSTA (without false path elimination). Out of this 19% reduction, 9% is due to the false path elimination (as observed from the results of SSTA with false path elimination in Table 8.5) and 10% is due to the use of different delay distributions for different input transitions (which cause a change in the output) for all gates in *LIB*. Therefore, to improve the accuracy of SSTA, it is important to consider both false paths in the circuit and also use different delay distributions for different input transitions at a gate (as done by StatSense).

8.4.1 Determining the Number of Input Vector Transitions N

The number of input vector transitions required to perform an accurate statistical timing analysis can be obtained as follows. Note that as the number of input vector transitions N is increased, the mean delay increases, while the standard deviation decreases. After a certain number N_1 of input vector transitions, the mean and the standard deviation of the delay will not change with an increase in the number of input vector transitions. This implies that when $N \geq N_1$, then all delay critical input vector transitions have already been considered and the new input vector transitions (the $(N_1 + 1)$th, $(N_1 + 2)$th, ... vectors) do not become ever critical under process variations. Therefore, N_1 input vector transitions are sufficient for an accurate statistical delay estimation. Although this method for calculating N was not used in this work, based on the results it is expected that N_1 is close to 75 for all the benchmark circuits analyzed in this work.

Table 8.5 Comparison of SSTA StatSense50, and SSTA without false paths

	SSTA			StatSense50				SSTA with false path elimination			
Ckt	μ (ps)	σ (ps)	$\mu + 3\sigma$	μ (ps)	σ (ps)	$\mu + 3\sigma$	Ratio	μ (ps)	σ (ps)	$\mu + 3\sigma$	Ratio
alu2	1, 008.39	19.08	1, 065.63	701.12	7.38	723.26	0.68	810.23	7.59	833	0.78
alu4	1,234.77	18.21	1,289.4	800.62	7.80	824.02	0.64	937.73	9.20	965.33	0.75
apex5	539.04	19.29	596.91	463.58	6.07	481.79	0.81	538.89	9.04	566.01	0.95
apex6	680.51	10.95	713.36	538.48	5.38	554.62	0.78	657.18	4.72	671.34	0.94
apex7	489.85	8.3	514.75	490.10	5.26	505.88	0.98	501.49	4.45	514.84	1.00
C499	737	11.29	770.87	650.35	6.18	668.89	0.87	748.65	5.85	766.2	0.99
C880	1,037.51	13.56	1,078.19	909.13	7.48	931.57	0.86	1, 047.92	7.23	1, 069.61	0.99
C1355	714.82	8.59	740.59	443.67	4.85	458.22	0.62	496.77	4.17	509.28	0.69
cordic	669.99	8.6	695.79	613.14	5.69	630.21	0.91	683.75	5.46	700.13	1.01
i6	496.16	22.8	564.56	493.86	8.39	519.03	0.92	537.32	10.64	569.24	1.01
i7	496.25	21.76	561.53	496.08	8.72	522.24	0.93	532.27	10.37	563.38	1.00
i10	1, 858.92	18.72	1,915.08	1,323.24	10.55	1,354.89	0.71	1,454.71	8.34	1,479.73	0.77
rot	781.23	13.75	822.48	540.15	7.33	562.14	0.68	656.28	5.37	672.39	0.82
x1	319.34	10.4	350.54	303.51	6.10	321.81	0.92	337.18	4.96	352.06	1.00
AVG							0.81				0.91

8.5 Chapter Summary

In recent times, the impact of process variations has become increasingly significant. Process variations have been growing larger and less systematic with each process generation. In response to this, there has been much research in extending traditional STA so that it can be performed statistically. The resulting SSTA approaches are, however, still quite pessimistic. This pessimism arises from the fact that most STA tools and their statistical counterparts do not consider false paths. The second major source of pessimism is that statistical static timing analyzers assume rising and falling delay distributions for all gates in a design to be a single Gaussian. However, the delay distribution of a gate is not necessarily Gaussian. In fact, the delay distribution for a multi-input gate is Gaussian for *each input vector transition* that causes a change on the gate output.

This chapter presented a sensitizable SSTA (which is referred to as StatSense) technique to overcome the pessimism of SSTA. The StatSense approach implicitly eliminates false paths, and also uses different delay distributions for the different input transitions of any gate. These features enable the StatSense approach to perform less conservative timing analysis than the SSTA approach. Experimental results show that on average, the worst case ($\mu + 3\sigma$) circuit delay reported by StatSense is about 19% lower than that reported by SSTA.

The next chapter describes a process variation tolerant combinational circuit design approach developed in this monograph.

References

1. M. Orshansky, S. R. Nassif, and D. Boning, *Design for manufacturability and statistical design: A constructive approach*, US Springer, 2008.
2. K. Agarwal and S. Nassif, "Characterizing process variation in nanometer CMOS," in *Proc. of the Design Automation Conf.*, June 2007, pp. 396–399.
3. K. Bernstein, D. J. Frank, A. E. Gattiker, W. Haensch, B. L. Ji, S. R. Nassif, E. J. Nowak, D. J. Pearson, and N. J. Rohrer, "High-performance CMOS variability in the 65-nm regime and beyond," *IBM Journal of Research and Development*, vol. 50, pp. 433–449, July/Sept. 2006.
4. H. Chang and S. S. Sapatnekar, "Statistical timing analysis under spatial correlations," *IEEE Transactions on Computer-Aided Design of Integrated Circuits and Systems*, vol. 24, no. 9, pp. 1467–1482, Sept. 2005.
5. J. Benkoski and A. J. Strojwas, "A new approach to hierarchical and statistical timing simulations," in *IEEE Transactions on Computer-Aided Design of Integrated Circuits and Systems*, Nov. 1987, vol. 6, pp. 1039–1052.
6. H. Jyu and S. Malik, "Statistical delay modeling in logic design and synthesis," in *Proc. of the Design Automation Conf.*, 1994, pp. 126–130.
7. C. Visweswariah, K. Ravindran, K. Kalafala, S. G. Walker, and S. Narayan, "First-order incremental block-based statistical timing analysis," in *Proc. of the Design Automation Conf.*, 2004, pp. 331–336.
8. A. Agarwal, V. Zolotov, and D. T. Blaauw, "Statistical timing analysis using bounds and selective enumeration," in *IEEE Transactions on Computer-Aided Design of Integrated Circuits and Systems*, Sept. 2003, vol. 22, pp. 1243–1260.

9. A. Agarwal, D. Blaauw, and V. Zolotov, "Statistical timing analysis for intra-die process variations with spatial correlations," in *Proc. of the Intl. Conf. on Computer-Aided Design*, Nov. 2003, pp. 900–907.
10. A. Agarwal, D. Blaauw, V. Zolotov, and S. Vrudhula, "Statistical timing analysis using bounds," in *Proc. of the Conf. on Design Automation and Test in Europe*, March.
11. A. Devgan and C. V. Kashyap, "Block-based static timing analysis with uncertainty," in *Proc. of the Intl. Conf. on Computer-Aided Design*, 2003, pp. 607–614.
12. J. Liou, K. Cheng, S. Kundu, and A. Krstic, "Fast statistical timing analysis by probabilistic event propagation," in *Proc. of the Design Automation Conf.*, 2001, pp. 661–666.
13. J. Liou, A. Krstic, L. Wang, and K. Cheng, "False-path-aware statistical timing analysis and efficient path selection for delay testing and timing validation," in *Proc. of the Design Automation Conf.*, 2002, pp. 566–569.
14. L. Xie and A. Davoodi, "Bound-based identification of timing-violating paths under variability," in *Proc. of the Asia and South Pacific Design Automation Conf.*, Jan. 2009, pp. 278–283.
15. P. McGeer, A. Saldanha, R. Brayton, and A. Sangiovanni-Vincentelli, *Logic Synthesis and Optimization*, chapter Delay Models and Exact Timing Analysis, pp. 167–189, US Kluwer Academic Publishers, Boston, MA, 1993.
16. E. M. Sentovich, K. J. Singh, L. Lavagno, C. Moon, R. Murgai, A. Saldanha, H. Savoj, P. R. Stephan, R. K. Brayton, and A. L. Sangiovanni-Vincentelli, "SIS: A system for sequential circuit synthesis," Tech. Rep. UCB/ERL M92/41, Electronics Research Laboratory, Univ. of California, Berkeley, May 1992.
17. S. A. Cook, "The complexity of theorem-proving procedures," in *Proc. of the Third Annual ACM Symposium on Theory of Computing*, 1971, pp. 151–158.
18. M. Davis, G. Logemann, and D. Loveland, "A machine program for theorem-proving," *Communication of the ACM*, vol. 5, no. 7, pp. 394–397, 1962.
19. S. Malik, Y. Zhao, C. F. Madigan, L. Zhang, and M. W. Moskewicz, "Chaff: Engineering an efficient SAT solver," in *Proc. of the Design Automation Conf.*, 2001, pp. 530–535.
20. Y. Kukimoto, W. Gosti, A. Saldanha, and R. K. Brayton, "Approximate timing analysis of combinational circuits under the XBD0 model," in *Proc. of the Intl. Conf. on Computer-Aided Design*, 1997, pp. 176–181.
21. L. Nagel, "Spice: A computer program to simulate computer circuits," in *University of California, Berkeley UCB/ERL Memo M520*, May 1995.
22. Y. Cao, T. Sato, D. Sylvester, M. Orshansky, and C. Hu, "New paradigm of predictive MOSFET and interconnect modeling for early circuit design," in *Proc. of IEEE Custom Integrated Circuit Conf.*, June 2000, pp. 201–204.
23. M. C. T. Chao, L. Wang, K. Cheng, and S. Kundu, "Static statistical timing analysis for latch-based pipeline designs," in *Proc. of the Intl. Conf. on Computer-Aided Design*, 2004, pp. 468–472.

Chapter 9
A Variation Tolerant Combinational Circuit Design Approach Using Parallel Gates

9.1 Introduction

The increasing variation in device parameters leads to large variations in the performance of different die of the same wafer, resulting in a significant yield loss. This yield loss translates into higher manufacturing costs. Therefore, it is important to design process variation tolerant circuits, to improve yield and lower manufacturing costs.

In this monograph, two design approaches are developed to improve the process variation tolerance of combinational circuits (described in this chapter) and voltage level shifters (discussed in the next chapter), respectively.

The process variation tolerant design approach for combinational circuits proposed in this chapter exploits the fact that random variations can cause a significant mismatch in the electrical performance of two identical devices placed next to each other on the die. In the proposed approach, a large gate is implemented using an appropriate number (>1) of smaller gates, whose inputs and outputs are connected to each other in parallel. This parallel connection of smaller gates to form a large gate is referred to as a parallel gate. Since the L and V_T variations are largely random (as discussed in Chap. 1) and have independent variations in the smaller gates, the variation tolerance of the parallel gate is improved. The parallel gates are implemented as single layout cells. By careful diffusion sharing in the layout of the parallel gates, it is possible to reduce the input and output capacitance of the gates, thereby improving the nominal circuit delay as well. An algorithm is also presented to selectively replace critical gates in a circuit by their parallel counterparts, to improve the variation tolerance of the circuit. Experiment results presented in Sect. 9.4 demonstrate that this process variation tolerant design approach achieves significant improvements in circuit level variation tolerance.

The rest of the chapter is organized as follows. Section 9.2 briefly discusses some previous work in this area. In Sect. 9.3, the proposed process variation tolerant design approach for combinational circuits is described. Experimental results are presented in Sect. 9.4, followed by the chapter summary in Sect. 9.5.

R. Garg and S. P. Khatri, *Analysis and Design of Resilient VLSI Circuits: Mitigating Soft Errors and Process Variations*, DOI 10.1007/978-1-4419-0931-2_9,

9.2 Related Previous Work

As mentioned in the previous chapter, process variation tolerant design has been an active research topic for several decades. Various approaches have been developed to efficiently analyze the effects of variations on the performance of a circuit [1,2,3,4] as well as to design process variation tolerant circuits [5,6,7,8].

To perform statistical circuit analysis and optimization, it is important to identify and characterize variation sources. Different circuit structures are reported [9,10,11, 12] to characterize and extract process variations (for both random and systematic variation components). In [10], significant variations were observed in the extracted threshold voltage values, and large mismatches were observed in adjacent SRAM devices fabricated in a 65 nm process. It was also argued that the large variation in V_T is mainly due to the random dopant fluctuations. The authors of [12] observed that the variations in L, V_T and mobility are major contributors to the overall variations in the performance of a circuit fabricated in a 65 nm SOI process. The variations in L and V_T were found to be normally distributed, with negligible spatial correlation [12]. This suggests that random variations are becoming more problematic than the systematic variations.

To evaluate the performance of a circuit under process variations, statistical timing analysis of a circuit is typically performed [1,2,3,4,13]. Some of the statistical timing analysis approaches have already been discussed in Chap. 8.

In [5], the authors perform gate sizing to improve the variation tolerance of digital circuits at the expense of an increase in the mean delay of the circuit. Thus, this approach does not improve the timing yield. A bidirectional adaptive body bias (ABB) technique is used to compensate for parameter variations in [14]. In this technique, a nonzero voltage is applied between the body and the source terminal to control the threshold voltage of transistors (and hence the speed of a circuit). In [15,6], the authors use both adaptive body bias (ABB) as well as adaptive supply voltage (ASV) to reduce the impact of the process variations. Using this technique, the number of die accepted in the highest three frequency bins increases to 98% [15] from 58%. Although ABB with ASV is very effective in improving yield, this technique can only be used to compensate for systematic variations. It is not feasible to apply a different body bias (and/or different supply voltages) to different gates in a circuit to compensate for random variations. Therefore, ABB (with or without ASV) cannot be used to deal with random variations. Since the variations of L and V_T are mostly random in nature, there is the need to develop techniques to reduce the impact of these random variations. Also, with diminishing feature sizes, the body effect coefficient is decreasing [16] and therefore, ABB-based approaches will not be effective for future technologies.

In [17], the authors present a defect tolerant design technique for nanodevices. In their approach, redundancy is added at the transistor level by replacing each transistor by a "quadded transistor." This is done to improve the functional reliability of a design against permanent defects such as stuck-open,stuck-shorts, and bridges.

Since each transistor in a design is replaced by 4 transistors, the area overhead of this approach is very large (>100% as reported in the paper). Another defect tolerant approach was presented in [18], where the authors duplicate transistors in a voter circuit to improve the functional reliability of triple modulo redundancy based fault-tolerant systems. The area overhead of this approach is also very high (~228%). The approaches of [17] and [18] try to improve only the *functional reliability* of a design at the *cost of area and delay overheads*. These approaches do not reduce the variability in the performance of the design, which is the goal of the work presented in this chapter. In contrast to these approaches, the proposed approach splits transistors to reduce *both* the mean and the standard deviation of the delay of a circuit.

9.3 Process Variation Tolerant Combinational Circuit Design

In Sect. 9.3.1, the variations considered in this chapter are described. Section 9.3.2 describes the proposed variation tolerant standard cell design approach. To improve the variation tolerance of a circuit, the gates in the circuit whose random variations result in a significant variability in the delay of the circuit are to be replaced by their variation tolerant counterparts. A circuit level approach proposed to improve variation tolerance is described in Sect. 9.3.3.

9.3.1 Process Variations

In this work, random variations in the L and V_T parameters of devices are considered, since these are the key parameters for determining the performance of a circuit. The authors of [12] extracted the variations in L and V_T for devices fabricated in a 65 nm SOI process. They found that the L and V_T of transistors are normally distributed and *vary independently*. The ratio of the standard deviation to the mean is 5% for L, and 9% for V_T. Based on this, both L and V_T of transistors are assumed to vary independently. Also, the standard deviation of L (σ_L) is taken to be 5% of its nominal value [12]. The standard deviation of the threshold voltage σ_{V_T} is a function of square root of the width (W) of a transistor, i.e., $\sigma_{V_T}(W) \propto 1/\sqrt{W}$ [19, 20, 21]. In [21], it is also reported that the σ_{V_T} of transistors varies with the channel width by a factor of at most 2. In other words, the σ_{V_T} of a very large device is approximately half of the σ_{V_T} of the smallest device [21]. This observation is based on extracted data from several test chips. Thus, the σ_{V_T} for the smallest device (with a width of $W = W_{min}$)[1] is taken to be 9% of the nominal threshold voltage value [12]. The largest σ_{V_T} for any device is taken to be 4.5%.

[1] The smallest device is a device with a width of 2× the feature size (or L_{min}). For a 65 nm process, the smallest device has a width $W_{min} = 130$ nm.

The σ_{V_T} for an arbitrary device with a width of W is obtained using the following relation:

$$\sigma_{V_T}(W) = \max\left\{\sigma_{V_T}(W_{\min})\sqrt{\frac{W_{\min}}{W}}, \frac{\sigma_{V_T}(W_{\min})}{2}\right\}. \quad (9.1)$$

9.3.2 Variation Tolerant Standard Cell Design

Random variations can cause a significant mismatch in the electrical performance of two identical devices placed next to each other. This phenomenon is utilized to design variation tolerant standard cells. Consider a 4× inverter shown in Fig. 9.1a. Assume that the transistor M1 (M2) of the 4× INV of Fig. 9.1a is implemented as a single NMOS (PMOS) transistor in the layout. This 4× INV is referred to as a regular inverter (an inverter implemented using a single PMOS and a single NMOS transistor). The L and V_T of M1 and M3 can vary randomly, which will directly affect the delay of the 4× INV, as well as the slew at the node *out*. This can also increase the delay variability of the circuit in which this inverter resides. To reduce the delay variability and the slew of the output of this INV due to random variations in L and V_T, the 4× INV is implemented by connecting two 2× inverters in parallel as shown in Fig. 9.1b. This implementation of the 4× INV (as shown in Fig. 9.1b) is referred as a "parallel" inverter. The parallel 4× INV is more tolerant to the random variations than a regular 4× INV since the variations in the L and V_T of transistors M1 and M3 are independent, and hence they tend to cancel each other. Similarly, the L and V_T variations of transistors M2 and M4 tend to cancel each other. Thus, the impact of random variations on the delay (and the slew) of the output of the parallel 4× INV is lower than that of the regular 4× INV. Monte-Carlo simulations were

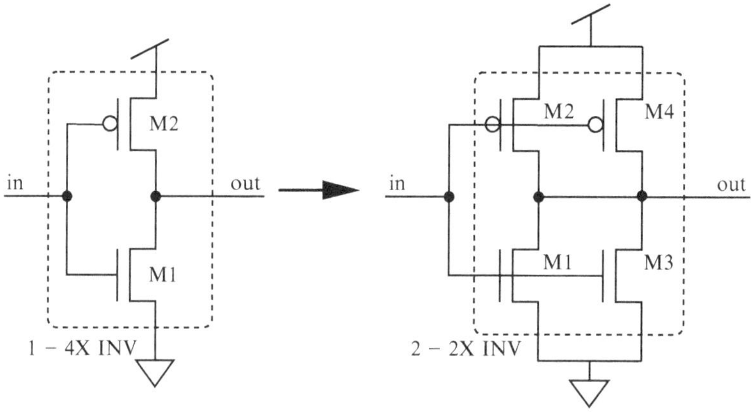

Fig. 9.1 4× inverter implementations

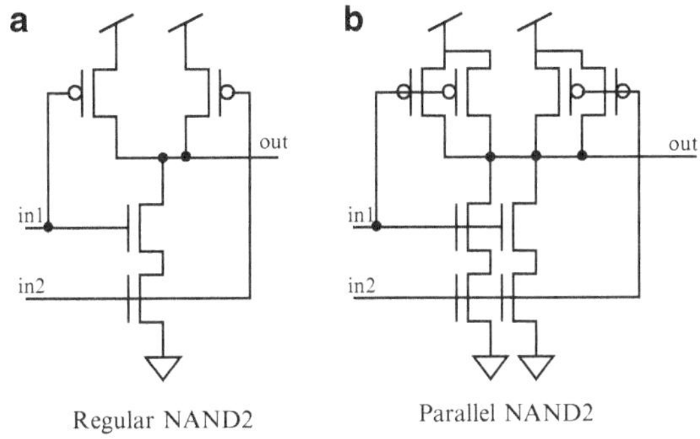

Fig. 9.2 2 input NAND gate: (**a**) regular, (**b**) parallel

performed to verify that the parallel 4× INV is more tolerant to random variation than the regular 4× INV. The results are presented in Sect. 9.4.

Using this approach, a variation tolerant complex gate G of size $k\times$ is designed by implementing it as a parallel connection of 2 (or more) smaller gates of size $k/2\times$ (or $k/3\times$, $k/4\times$, ...), with the same functionality as the gate G. In other words, instead of using large PMOS and large NMOS transistors to implement G, small PMOS and NMOS transistors are used, and connected in parallel to improve the variation tolerance of G. Figure 9.2 shows the regular and the variation tolerant (parallel) versions of a 2-input NAND gate. Note that the number of small transistors that can be connected in parallel to implement a large transistor (inside a gate) is constrained by the width of the smallest device that can be fabricated. This fact is taken into consideration while designing variation tolerant parallel gates. For example, in a regular NOR2 gate of minimum size, both NMOS transistors are of minimum width. Therefore, in the parallel version of this gate, there are still 2 minimum width NMOS transistors (connected to the two inputs). Since the PMOS transistors of the regular NOR2 gates are 5× devices, they can be implemented using smaller PMOS transistors connected in parallel. Layouts were created for regular and parallel versions of all gates in the library (*LIB*) used in this work. The parallel gates were implemented as single layout cells. The same cell height was used for the regular and the parallel versions of any gate. This permits seamless placement and routing of a circuit using both regular as well as parallel gates. Since the parallel gates utilize more transistors than their regular counterparts and both have the same height, the layout area of a parallel gate is more than that of the corresponding regular gate (since a larger cell width is needed to include more transistors in the parallel gate). To limit the area overhead of the proposed approach, regular gates in a circuit are selectively replaced by parallel gates to improve the variation tolerance of the circuit. The approach used to select regular gates to be replaced is explained in Sect. 9.3.3.

Apart from increasing the variation tolerance of a gate, another advantage of the approach proposed in this chapter is that the input capacitance (C_{in}) of any pin of a parallel gate G is lower than the input capacitance of the corresponding pin

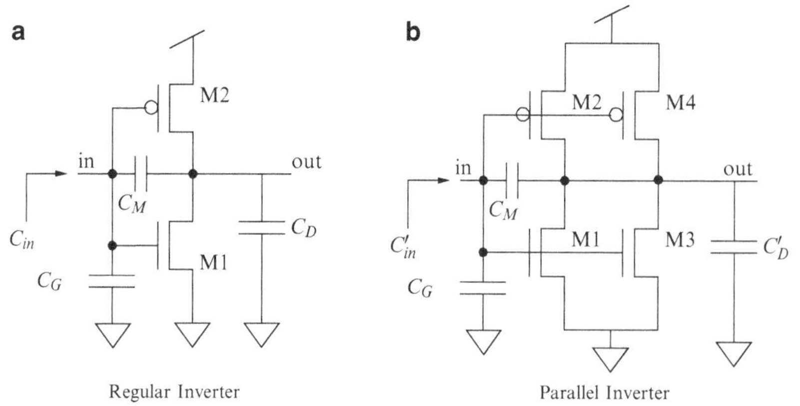

Fig. 9.3 Capacitance of various nodes: (**a**) regular inverter, (**b**) parallel inverter

of the regular gate. This is explained next. Consider a regular and a parallel inverter of equivalent size, shown in Fig. 9.3. Figure 9.3 also shows the capacitance at the input and the output nodes of both inverters. The capacitance C_{G} is summation of the gate capacitance of the transistors M1 and M2 (M1, M2, M3, and M4) of the regular (parallel) inverter. C_{D} (C'_{D}) is the total output diffusion capacitance of the regular (parallel) inverter. C_{M} is the Miller capacitance between the input and the output of both inverters. The parallel inverter has two PMOS (NMOS) transistors in parallel. Therefore, in the layout of this inverter, transistors M2 and M4 will share their diffusion. Similarly, M1 and M3 will also share their drain diffusion. Thus, the total area of the output diffusion node is lower in the parallel inverter compared with the regular inverter. This implies that $C'_{\mathrm{D}} < C_{\mathrm{D}}$. Note that C_{G} and C_{M} are identical for both regular and parallel inverters, since the total width of PMOS and NMOS devices is equal in both these inverters. The input capacitance of the regular (parallel) inverter C_{in} (C'_{in}) depends on C_{G}, C_{M}, and C_{D} (C'_{D}). In particular, $C_{\mathrm{in}} = C_{\mathrm{G}} + C_{\mathrm{D}}C_{\mathrm{M}}/(C_{\mathrm{D}} + C_{\mathrm{M}})$ and $C'_{\mathrm{in}} = C_{\mathrm{G}} + C'_{\mathrm{D}}C_{\mathrm{M}}/(C'_{\mathrm{D}} + C_{\mathrm{M}})$. The input capacitance of the parallel inverter is thus lower than the input capacitance of the regular inverter since $C'_{\mathrm{D}} < C_{\mathrm{D}}$. Note that since $C'_{\mathrm{D}} < C_{\mathrm{D}}$, the intrinsic delay is also lower for the parallel inverter compared with the regular inverter. The lower input capacitance and the lower intrinsic delay of the parallel gates helps in reducing the circuit level delay. Thus, the use of parallel gates in a circuit (instead of regular gates) can reduce both the mean (μ) *and* the standard deviation (σ) of the delay of the circuit. The delay limited yield also improves since the worst case circuit delay ($\mu + 3\sigma$) decreases. This is demonstrated for several benchmark circuits in Sect. 9.4. Another advantage of using parallel gates in a circuit is that the dynamic power consumption of the circuit reduces because of the lower input and output capacitances of the parallel gates.

9.3.3 Variation Tolerant Combinational Circuits

As mentioned earlier, the layout area of parallel gates is higher than that of regular gates. Therefore, replacing regular gates with parallel gates in a circuit (to improve its tolerance to random variations) would incur an area penalty. To minimize this area penalty, only those gates that contribute significantly to the performance variability of the circuit are replaced by their variation tolerant parallel counterparts. In this work, such gates are identified based on their deterministic slack value. It is reasonable to expect that delay variations of a gate with a low slack value are likely to induce significant variations in the performance of the circuit. Therefore, regular gates in a circuit are replaced by parallel gates in increasing order of their slack value. The number of gates in a circuit that will be replaced depends on a user specified area constraint. A user specified number P, which is the fraction of total number of gates N in a circuit to be replaced with their variation tolerant counterparts, is used in the proposed approach. This number P can be obtained from the amount of area penalty that is acceptable.

First, the slack is computed for all the gates in the circuit (η) and then the critical gates are identified and replaced using Algorithm 4. For a given circuit η, first sort all gates $G \in \eta$ in an increasing order of their slack values, and store them in a list L. Then replace the top $P * N$ gates (by replacing them with their corresponding variation tolerant parallel gates) in the list L. The resulting variation tolerant circuit is referred to as η^*. This approach is referred to as the *percentage* gate replacement approach.

Algorithm 4 Replacing critical gates in η to improve its variation tolerance

```
Increase_variation_tolerance(η, P)
L = sort(G ∈ η, Slack(G))
i = 0
N = number_of_gates(η)
while i < P * N do
   G = L(i)
   Replace G by G_parallel
   i = i + 1
end while
return η*
```

The identification of critical gates (under process variations) for circuit optimization is an active research topic [22, 23, 24, 25, 26]. To keep the proposed approach efficient, the deterministic slack value is used to identify critical gates in a circuit. However, it is possible to use the approach of [22] to identify gates in a circuit, which are critical under variations and replace them with the proposed variation tolerant parallel gates. Using the approach of [22], the area overhead of the proposed approach may be lowered. It is also possible to improve the reduction in the delay variability of a circuit.

9.4 Experimental Results

Monte-Carlo simulations of the 4× regular and parallel INVs of Fig. 9.1 were performed using SPICE [27]. A 65 nm PTM [28] model card was used, with $VDD = 1$ V. The L and V_T of the different transistors of these inverters were varied independently. Also, σ_L was taken to be 5% of its nominal value [12] and σ_{V_T} was computed using the method described in Sect. 9.3.1. Figure 9.4a shows the standard deviation of the rising and the falling delay of both the regular and the parallel 4× INVs, for different loads (assuming an input slew of 30 ps). Figure 9.4b compares the worst case ($\mu + 3\sigma$) rising and falling delays and Figure 9.4c plots the standard deviation of the output slew. Figure 9.4 clearly shows that the parallel 4× INV is less impacted by random variations in the L and V_T of devices compared with the regular 4× INV. The worst case rising and falling delays of the parallel INV are lower than those of the regular inverter by ∼8% and ∼4%, respectively. Therefore, the parallel INV of Fig. 9.1 is more tolerant to random variations compared with the regular inverter.

A standard cell library (*LIB*) was implemented using a 65 nm PTM [28] model card, with VDD = 1.0 V. The standard cell library *LIB* consists of regular INV2X,

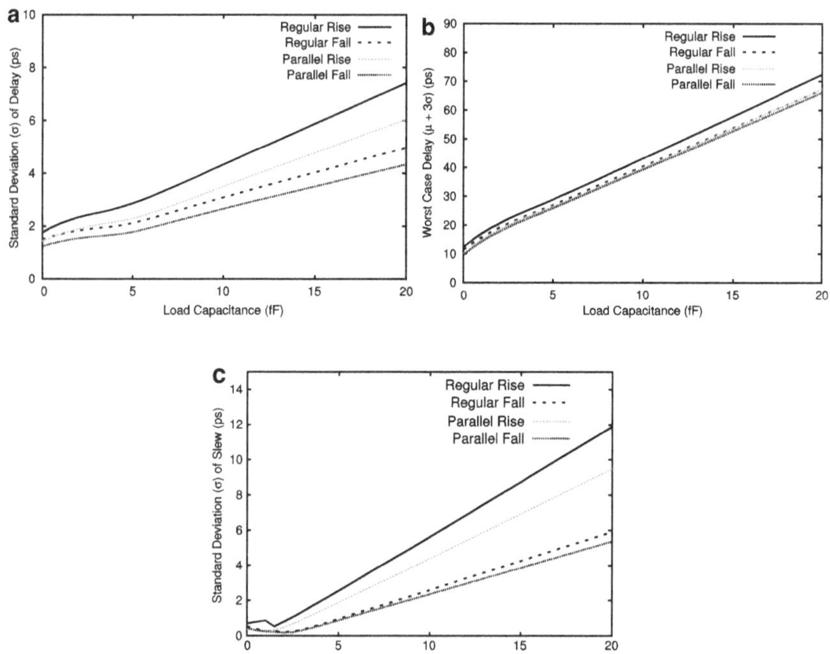

Fig. 9.4 Results for 4× regular and parallel inverters: (**a**) standard deviation of delay, (**b**) worst case delay, and (**c**) standard deviation of output slew

INV4X, INV8X, NAND2X2, NAND2X4, NAND3X2, NOR2X2, and NOR3X2 gates[2]. The variation tolerant (parallel) versions of all the gates in *LIB* were also designed. For regular INV2X, INV4X, NAND2X2, NAND3X2, NOR2X2, and NOR3X2 gates, one parallel version (INV2XP, INV4XP, NAND2X2P, NAND3X2P, NOR2X2P, and NOR3X2P) was created. Two parallel versions were created for INV8X and NAND2X4. INV8X can be implemented as either 2 INV4X's in parallel (INV8XP1) or 4 INV2X's in parallel (INV8XP2). Similarly, the parallel versions of NAND2X4 are NAND2X4P1 and NAND2X4P2, where NAND2X4P2 utilizes more NMOS and PMOS transistors connected in parallel than NAND2X4P1. The minimum width of a transistor that can be fabricated is a 65 nm process is 130 nm. This was taken into account while creating variation tolerant (parallel) gates. Note that in some gates (INV2X, NOR2X2, and NOR3X2), only the PMOS devices could be parallelized since the NMOS devices were minimum sized. Layouts were created for all regular and parallel gates using CADENCE SEDSM [29] tools, paying careful attention to invoke diffusion sharing whenever feasible.

The standard cells in *LIB* (both regular and parallel versions) were characterized to construct 2-D lookup tables for the values of the mean and standard deviation of the input pin to output delay, as well as the output slew. This precharacterization was done for a set of load capacitance and input slew values. Table 9.1 compares the mean (μ), the standard deviation (σ), and the worst case delay ($\mu + 3\sigma$) of all regular gates in *LIB* along with their variation tolerant counterparts, for a load value of 5 fF and an input slew of 30 ps. In this table, the INV8X and NAND2X4 gates are compared with the INV8XP2 and NAND2X4P2 gates, respectively. For multiple input gates, the input pin with the largest mean pin to output delay value was used for this comparison. Table 9.1 also compares the layout area, and the mean and standard deviation of the subthreshold leakage current of the regular and the parallel gates. The subthreshold leakage of a gate is obtained for the input state which maximizes its value. In Table 9.1, Column 1 reports gates under consideration. Columns 2–10 report the results obtained for the regular gates in *LIB*. Columns 11–19 report the ratio of any quantity for the parallel gate compared with the value of the same quantity for the regular gate. For example, Column 9 reports the ratio of the mean rising delay of a parallel gate to the mean rising delay of the corresponding regular gate. As reported in Table 9.1, on average the standard deviation of the rising (falling) delay of the parallel gates is lower by 31% (15%) compared with the regular gates. Also, the mean and the worst case rising (falling) delays of the parallel gates are lower by 2% and 10% (1% and 4%), respectively, compared with the regular versions. Hence, the proposed parallel gates are more tolerant to random variations than the regular gates. The average layout area of the parallel gates is higher by 60% compared with regular gates. The average (over all parallel gates) of the mean (under process variations) subthreshold leakage current is 1.01× that of the regular gates. However, the average (over all parallel gates) of the standard

[2] INV2X is the smallest inverter that can be manufactured. The width of the NMOS (PMOS) transistor of INV2X is 130 nm (325 nm).

Table 9.1 Comparison of regular and parallel gates

	Regular gates									Parallel gates								
	Rising delay (ps)			Falling delay (ps)			Area	Leakage (nA)		Rising delay ratio			Falling delay ratio				Leakage ratio	
Gate	μ	σ	$\mu + 3\sigma$	μ	σ	$\mu + 3\sigma$	(μm^2)	μ	σ	μ	σ	$\mu + 3\sigma$	μ	σ	$\mu + 3\sigma$	Area ratio	μ	σ
INV2X	30.72	4.91	45.44	32.42	3.43	42.71	1.62	67.96	418.17	0.99	0.79	0.92	0.98	0.99	0.98	1.33	1.05	0.73
INV4X	20.24	2.85	28.80	20.59	2.12	26.95	1.62	130.70	762.48	0.97	0.81	0.92	0.99	0.84	0.96	1.33	0.92	0.64
INV8X	14.16	2.56	21.85	14.75	1.60	19.55	2.16	252.18	1,622.38	0.98	0.55	0.83	0.98	0.65	0.90	1.50	1.04	0.53
NAND2X2	34.59	5.79	51.95	32.11	1.99	38.09	2.16	122.80	463.55	0.99	0.66	0.88	1.00	0.82	0.97	1.50	1.09	0.89
NAND2X4	24.13	3.46	34.51	22.65	1.13	26.05	2.16	257.26	1,081.43	0.98	0.56	0.85	1.01	0.67	0.96	2.50	1.10	0.58
NAND3X2	40.81	6.86	61.40	34.81	1.56	39.50	2.70	204.50	700.15	0.97	0.74	0.89	0.98	0.83	0.97	1.60	1.02	0.74
NOR2X2	37.64	3.50	48.14	40.82	4.35	53.87	2.16	131.79	639.64	0.98	0.75	0.93	0.98	1.01	0.99	1.50	0.87	0.66
NOR3X2	46.50	3.68	57.54	54.50	5.82	71.97	2.70	157.04	708.34	0.99	0.69	0.93	0.97	0.97	0.97	1.60	0.97	0.71
AVG										0.98	0.69	0.90	0.99	0.85	0.96	1.61	1.01	0.69

deviation of the subthreshold leakage current is 31% lower compared with the same quantity for regular gates. Thus, the approach proposed in this chapter also reduces the variability in the subthreshold leakage current, with a small increase in its mean value. Note that the input pin capacitance and the output capacitance of the parallel gates are smaller by 2.5% and 15% on average, compared with the corresponding capacitances of the regular gates. The improvements in the delay μ, σ, and $\mu + 3\sigma$ are higher for rising transitions, since there are more opportunities to parallelize PMOS devices since they are nominally larger than the NMOS devices.

Several ISCAS and MCNC benchmark circuits were mapped using *LIB*, for both area and delay optimality. For a mapped design, the slack of all the gates in the design was computed, and then the gates were sorted in order of increasing slack. After this, the gates with the lowest slack in the design were replaced by their variation tolerant counterparts using *percentage* based gate replacement approach described in Sect. 9.3.3 (until a fraction P of the total number of gates in the design are replaced). For both regular area and delay mapped designs (mapping was performed in SIS [30]), two variation tolerant versions were generated using parallel gates. In the first version, the INV8XP1 and NAND2X4P1 parallel versions for regular INV8X and NAND2X4 gates were used, respectively (for all other regular gates, note that there is only parallel version). The resulting area and delay mapped designs are referred to as AM1 and DM1 designs, respectively. In the second version of variation tolerant circuits, the INV8XP2 and NAND2X4P2 parallel versions were used for regular INV8X and NAND2X4. The resulting area and delay mapped designs are referred to as AM2 and DM2, respectively. Note that since INV8X and NAND2X4 have a large area, on the one hand, the area mapped designs did not utilize these gates. Hence, the AM1 and AM2 results are identical; therefore, the results are presented only for AM1. On the other hand, delay mapped designs use these large gates heavily, and thus significant differences exist between DM1 and DM2. The results are presented for both DM1 and DM2.

For regular area and delay mapped designs as well as for variation tolerant circuits AM1, DM1, and DM2, Monte-Carlo-based statistical static timing analysis (SSTA) was performed to obtain their delay distributions. Monte-Carlo-based SSTA is considered to be an accurate method to obtain the delay distribution of a circuit [31]. Monte-Carlo-based SSTA was implemented in *SIS* [30]. These data obtained from the characterization of all standard cells (regular and parallel versions) in the library (*LIB*) were used for Monte-Carlo-based SSTA. Note that the characterization of all gates was done for a set of different load capacitance and input slew values. Ten thousand iterations were performed for the Monte-Carlo-based SSTA analysis of any circuit. The mean (μ), the standard deviation (σ), and the worst case delay ($\mu + 3\sigma$) of all regular and variation tolerant designs were obtained from this SSTA analysis. The layout area of all designs was also computed. Note that the area for a design was computed by adding the layout area of all the gates in the circuit. Figure 9.5 shows the average (over 14 benchmark designs) ratio of the area and delay results of the variation tolerant circuits compared with their regular counterparts, for different varying values of P from (0 to 1). Note that P is a user specified number, which is the fraction of total number of gates in a circuit to be

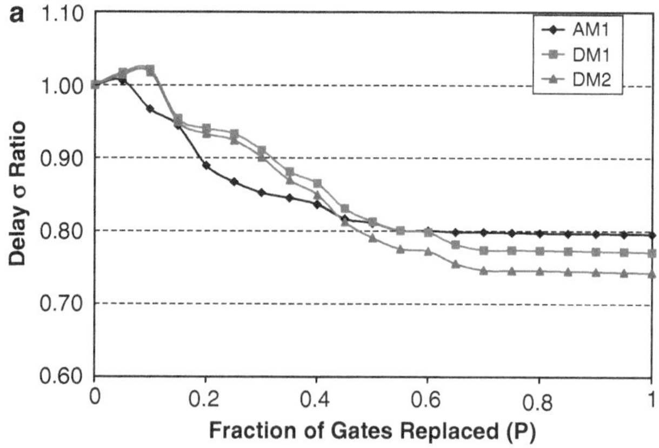

Delay σRatio

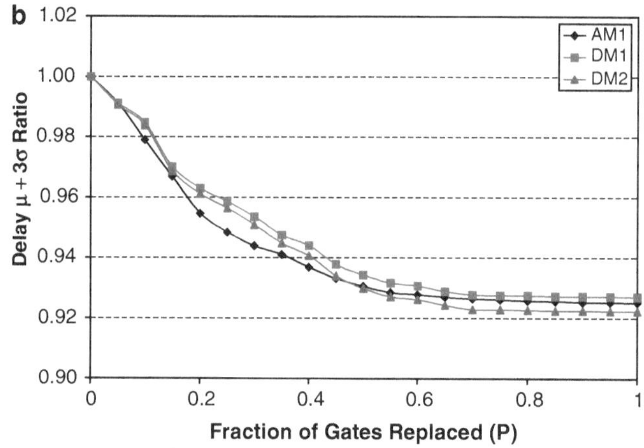

Delay $\mu + 3\sigma$ Ratio

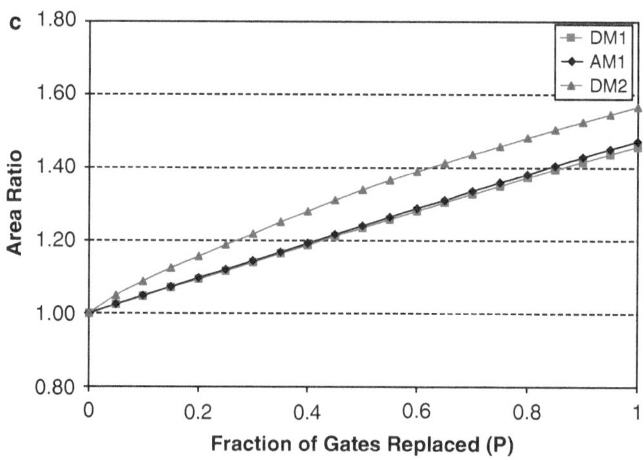

Area Ratio

Fig. 9.5 Ratio of results of the proposed approach compared with regular circuits for different values of P

replaced with their parallel versions. Figures 9.5a, b plots the average (averaged over all benchmark circuits) ratio of σ and $\mu + 3\sigma$ of the delay of the variation tolerant circuits (AM1, DM1 and DM2) compared with regular designs. Figure 9.5c shows the average ratio of the layout area of the variation tolerant design compared with the regular version (as a function of P). As shown in as Fig. 9.5a, b, on average, both the σ and $\mu + 3\sigma$ of the delay of the AM1, DM1, and DM2 designs reduce as the fraction of parallel gates (P) in the designs increases. When P reaches 0.7, the σ and $\mu + 3\sigma$ of the delay of the variation tolerant designs saturate. At this point, on average, the σ and $\mu + 3\sigma$ of AM1 is 20% and 7% lower than that of regular area mapped designs. The area utilization of AM1 is around 34% more than regular designs. For DM1 (DM2), at $P \cong 0.7$, the σ and $\mu + 3\sigma$ are 23% and 7% (25% and 8%) lower than regular delay mapped designs, with an area penalty of 33% (44%). From this it can be concluded that the variation tolerant design approach presented in this chapter reduces both the σ and $\mu + 3\sigma$ of the delay of designs significantly and hence increases the delay limited design yield. Also, the DM2 designs perform better than DM1 on average but with a higher area penalty. Although this is not shown explicitly, the mean delays of the AM1, DM1, and DM2 designs are also lower compared with regular area and delay mapped designs (by 6%, 6% and 6%, respectively). The mean delays of AM1, DM1, and DM2 designs are lower because the input pin capacitance and the output capacitance of the parallel gates are smaller (as explained in Sect. 9.3.2) by 2.5% and 15% on average, compared with the corresponding capacitances of the regular gates.

The σ and $\mu + 3\sigma$ of delay as well as the area ratio of individual AM1 designs compared with their corresponding regular version are plotted in Fig. 9.6, for different values of P. Notice from Fig. 9.6 that for smaller values of P, the reduction in σ and $\mu + 3\sigma$ for some benchmark circuits (with increasing P) is abrupt. When P reaches 0.6, then the σ and $\mu + 3\sigma$ of all benchmark circuits either saturate or decrease very slowly with increasing P. Therefore, $P = 0.7$ is a reasonable value to be used in the *percentage*-based gate replacement approach. As expected, the area of AM1 designs increase linearly with increasing P. Similar trends are also observed for DM1 and DM2. The σ, $\mu + 3\sigma$, and area ratio plots for individual DM1 and DM2 designs are shown in Figs. 9.7 and 9.8.

From Table 9.1, it can be concluded that the mean, the standard deviation, and the worst case delay of the variation tolerant (parallel) gates are lower than that of their regular counterparts. At the circuit level, as shown in Fig. 9.5, both the σ and $\mu + 3\sigma$ of variation tolerant circuits (obtained by using parallel gates) are lower that of the regular designs. Note that it is possible to use the approach of [22] to identify the critical gates in a circuit under variations and possibly reduce the area overhead of the approach presented in this chapter.

The effect of implementing a large transistor by a parallel connection of small transistors, for a large 32× INV was also studied in this work. For this study, only the V_T of transistors was varied, and Monte-Carlo simulations of different implementations of the 32× INV were performed. The different implementations of 32× are as follows: 1-32×, 2-16×, 4-8×, 8-4× and 16-2×. The results of these simulations show that initially, the σ of delay of these implementations decreases as the

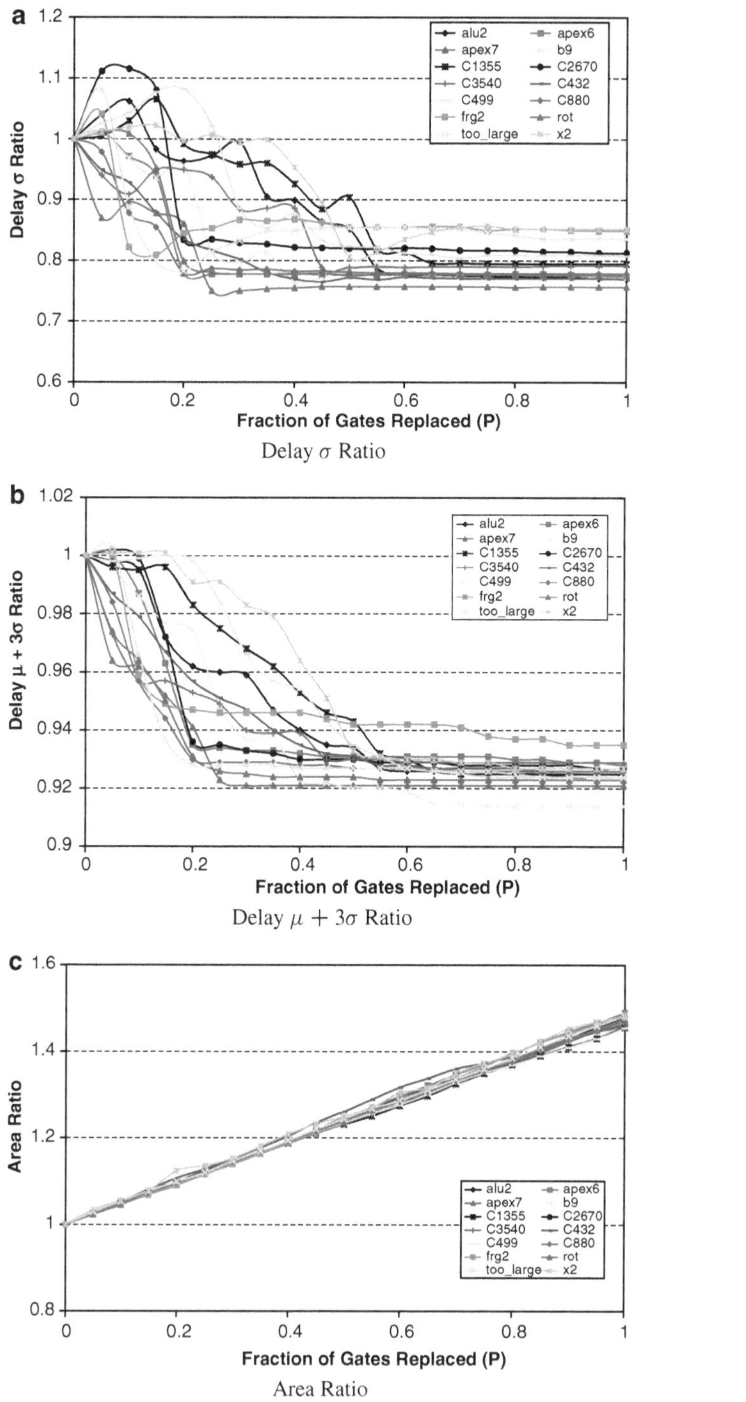

Delay σ Ratio

Delay $\mu + 3\sigma$ Ratio

Area Ratio

Fig. 9.6 Delay σ, $\mu + 3\sigma$, and area ratio of the proposed approach compared with regular circuits for different values of P for area mapped (AM1) designs

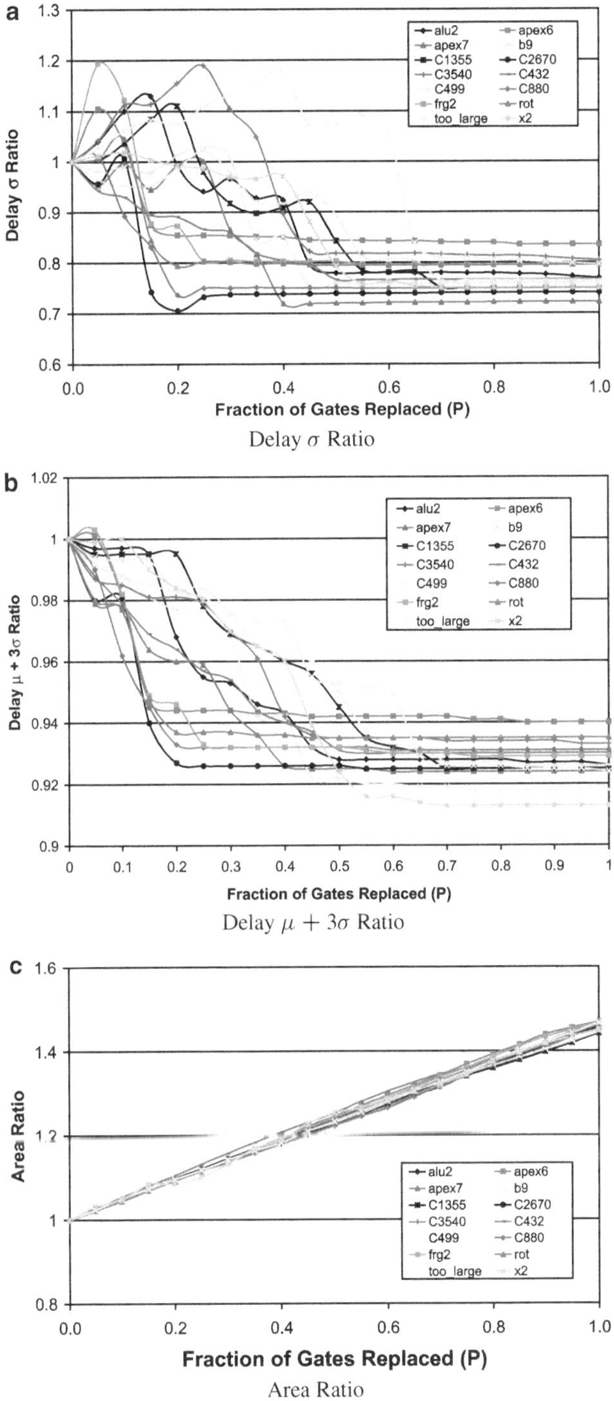

Fig. 9.7 Delay σ, $\mu + 3\sigma$ and area ratio of the proposed approach compared with regular circuits for different values of P for DM1 designs

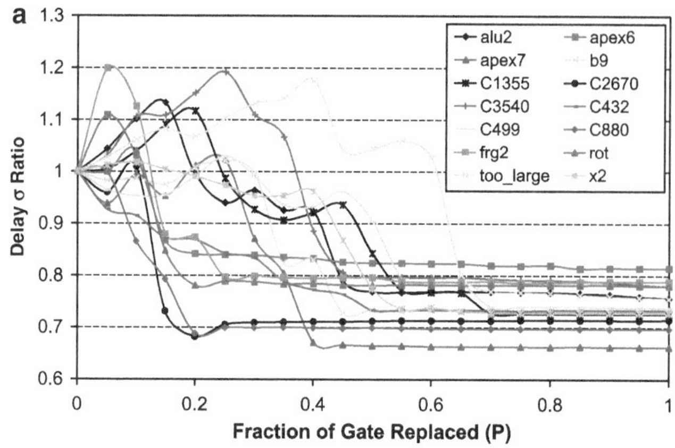

Delay σ Ratio

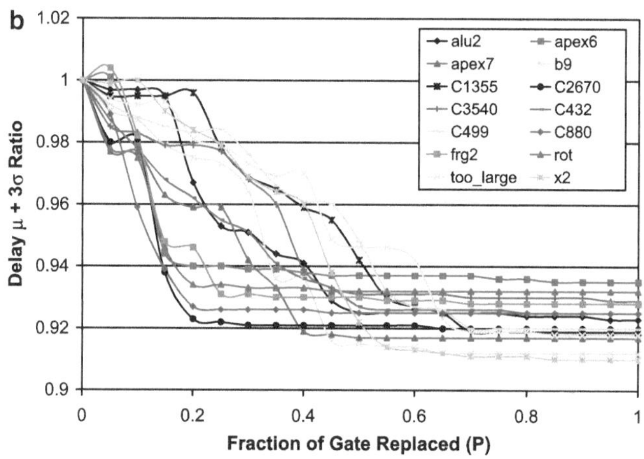

Delay $\mu + 3\sigma$ Ratio

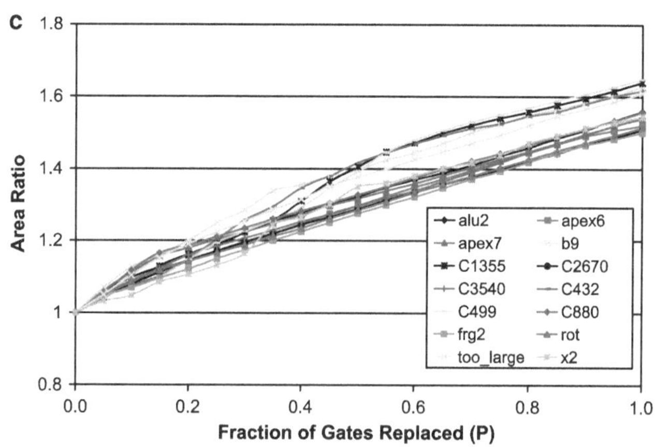

Area Ratio

Fig. 9.8 Delay σ, $\mu + 3\sigma$, and area ratio of the proposed approach compared with regular circuits for different values of P for DM2 designs

number of smaller inverters connected in parallel increases (until 8-4×) and then it saturates. The mean delay also goes down initially (until 4-8×) due to a reduction in the diffusion area, and then it starts increasing. When a large number of smaller inverters are used to implement a 32× INV, the perimeter capacitance of the output diffusion node dominates the bottom plate capacitance. As a result, the total capacitance at the output node of the 32× INV starts increasing with an increase in the number of smaller inverters used to implement it. Therefore, a large gate should be implemented using a parallel connection of an appropriate number (typically 2-4) of smaller gates. Using too many smaller gates to implement a large gate may increase the mean delay and hence affect the delay limited yield.

9.5 Chapter Summary

With the continuous scaling of devices, variations in key device parameters such as channel length (L), threshold voltage (V_T), and oxide thickness (T_{ox}) are increasing at an alarming rate. This has led to significant problems in terms of reliability, circuit resilience, and yields. In this chapter, a circuit design approach was proposed and validated, to alleviate this problem for combinational circuits. The proposed approach implements a large gate using an appropriate number (>1) of smaller gates connected in parallel (with their inputs and outputs connected to each other). Since the L and V_T variations are largely random and vary independently in the smaller gates, the variation tolerance of the parallel gates is improved. The parallel gates were implemented as single layout cells, and have smaller delay μ and σ compared with their traditional counterparts. This chapter also presented an algorithm to selectively replace critical gates in a circuit by their parallel counterparts, to improve circuit-level variation tolerance. Monte-Carlo simulations demonstrate that the proposed variation tolerant circuit design approach achieves significant improvements in delay σ and $\mu + 3\sigma$ variation. On average, the proposed approach reduces the standard deviation (σ) of the circuit delay by 23% for delay mapped designs, with an area overhead of 33% (compared with regular circuits). This approach also reduces the worst case circuit delay under variations (i.e., $\mu + 3\sigma$) by 7% and hence significantly improves the design yield.

References

1. H. Chang and S. S. Sapatnekar, "Statistical timing analysis under spatial correlations," *IEEE Transactions on Computer-Aided Design of Integrated Circuits and Systems*, vol. 24, no. 9, pp. 1467–1482, Sept. 2005.
2. C. Visweswariah, K. Ravindran, K. Kalafala, S. G. Walker, and S. Narayan, "First-order incremental block-based statistical timing analysis," in *Proc. of the Design Automation Conf.*, 2004, pp. 331–336.

3. A. Agarwal, D. Blaauw, and V. Zolotov, "Statistical timing analysis for intra-die process variations with spatial correlations," in *Proc. of the Intl. Conf. on Computer-Aided Design*, Nov. 2003, pp. 900–907.
4. A. Agarwal, D. Blaauw, V. Zolotov, and S. Vrudhula, "Statistical timing analysis using bounds," in *Proc. of the Conf. on Design Automation and Test in Europe*, March.
5. O. Neiroukh and X. Song, "Improving the process-variation tolerance of digital circuits using gate sizing and statistical techniques," in *Proc. of the Conf. on Design Automation and Test in Europe*, 2005, pp. 294–299.
6. J. Tschanz, K. Bowman, and V. De, "Variation-tolerant circuits: Circuit solutions and techniques," in *Proc. of the Design Automation Conf.*, June 2005, pp. 762–763.
7. G. Nabaa and F. N. Najm, "Minimization of delay sensitivity to process induced voltage threshold variations," in *Proc. of the IEEE-NEWCAS Conf.*, June 2005, pp. 171–174.
8. S. Bhunia, S. Mukhopadhyay, and K. Roy, "Process variations and process-tolerant design," in *Proc. of the Intl. Conf. on VLSI Design*, Jan. 2007, pp. 699–704.
9. A. Gattiker, M. Bhushan, and M. B. Ketchen, "Data analysis techniques for CMOS technology characterization and product impact assessment," in *Proc. of the Intl. Test Conf.*, 2006, pp. 1–10.
10. K. Agarwal, F. Liu, C. McDowell, S. Nassif, K. Nowka, M. Palmer, D. Acharyya, and J. Plusquellic, "A test structure for characterizing local device mismatches," in *Proc. of the Symposium on VLSI Circuits*, 2006, pp. 67–68.
11. K. Agarwal and S. Nassif, "Characterizing process variation in nanometer CMOS," in *Proc. of the Design Automation Conf.*, June 2007, pp. 396–399.
12. W. Zhao, Y. Cao, F. Liu, K. Agarwal, D. Acharyya, S. Nassif, and K. Nowka, "Rigorous extraction of process variations for 65nm CMOS design," in *Proc. of the European Solid State Device Research Conf.*, Sept. 2007.
13. R. Garg, N. Jayakumar, and S. P. Khatri, "On the improvement of statistical timing analysis," in *Proc. of the Intl. Conf. on Computer Design*, Oct. 2006, pp. 37–42.
14. J. Tschanz, J. Kao, S. Narendra, R. Nair, D. Antoniadis, A. Chandrakasan, and Vivek De, "Adaptive body bias for reducing impacts of die-to-die and within-die parameter variations on microprocessor frequency and leakage," in *Proc. of the Intl. Solid State Circuits Conf.*, Feb. 2002, pp. 422 – 478.
15. J.W. Tschanz, S. Narendra, R. Nair, and V. De, "Effectiveness of adaptive supply voltage and body bias for reducing impact of. parameter variations in low power and high performance microprocessors," *IEEE Journal of Solid-State Circuits*, vol. 38, no. 5, pp. 826–829, May 2003.
16. B. C. Paul, A. Agarwal, and K. Roy, "Low-power design techniques for scaled technologies," *Integration, the VLSI Journal*, vol. 39, no. 2, pp. 64 – 89, 2006.
17. A. H. El-Maleh, B. M. Al-Hashimi, and A. Melouki, "Transistor-level based defect tolerance for reliable nanoelectronics," in *Proc. of the Intl. Conf. on Computer Systems and Applications*, April 2008, pp. 53–60.
18. H. El-Razouk and Z. Abid, "A new transistor-redundant voter for defect-tolerant digital circuits," *Proc. of the Canadian Conf. on Electrical and Computer Engineering*, pp. 1078–1081, May 2006.
19. M. J. M. Pelgrom, A. C. J. Duinmaijer, and A. P. G. Welbers, "Matching properties of MOS transistors," *IEEE Journal of Solid-State Circuits*, vol. 24, no. 5, pp. 1433–1439, Oct 1989.
20. M. Orshansky, S. R. Nassif, and D. Boning, *Design for manufacturability and statistical design: A constructive approach*, US Springer, 2008.
21. K. Bernstein, D. J. Frank, A. E. Gattiker, W. Haensch, B. L. Ji, S. R. Nassif, E. J. Nowak, D. J. Pearson, and N. J. Rohrer, "High-performance CMOS variability in the 65-nm regime and beyond," *IBM Journal of Research and Development*, vol. 50, pp. 433–449, July/Sept. 2006.
22. M. R. Guthaus, N. Venkateswarant, C. Visweswariaht, and V. Zolotov, "Gate sizing using incremental parameterized statistical timing analysis," in *Proc. of the Intl. Conf. on Computer-Aided Design*, 2005, pp. 1029–1036.
23. S. Raj, S. B. K. Vrudhula, and J. Wang, "A methodology to improve timing yield in the presence of process variations," in *Proc. of the Design Automation Conf.*, 2004, pp. 448–453.

24. A. Agarwal, K. Chopra, and D. Blaauw, "Statistical timing based optimization using gate sizing," in *Proc. of the Conf. on Design Automation and Test in Europe*, 2005, pp. 400–405.
25. X. Bai, C. Visweswariah, and P. N. Strenski, "Uncertainty-aware circuit optimization," in *Proc. of the Design Automation Conf.*, 2002, pp. 58–63.
26. S. H. Choi, B. C. Paul, and K. Roy, "Novel sizing algorithm for yield improvement under process variation in nanometer technology," in *Proc. of the Design Automation Conf.*, 2004, pp. 454–459.
27. L. Nagel, "Spice: A computer program to simulate computer circuits," in *University of California, Berkeley UCB/ERL Memo M520*, May 1995.
28. Nanoscale integration and modeling (NIMO) group (2007), *ASU Predictive Technology Model [On-line]*, Available: http://www.eas.asu.edu/~ptm.
29. Cadence Design Systems, Inc., San Jose, CA, *Envisia Silicon Ensemble Place-and-route Reference Manuals*, Nov. 1999.
30. E. M. Sentovich, K. J. Singh, L. Lavagno, C. Moon, R. Murgai, A. Saldanha, H. Savoj, P. R. Stephan, R. K. Brayton, and A. L. Sangiovanni-Vincentelli, "SIS: A system for sequential circuit synthesis," Tech. Rep. UCB/ERL M92/41, Electronics Research Laboratory, Univ. of California, Berkeley, May 1992.
31. M. C. T. Chao, L. Wang, K. Cheng, and S. Kundu, "Static statistical timing analysis for latch-based pipeline designs," in *Proc. of the Intl. Conf. on Computer-Aided Design*, 2004, pp. 468–472.

Chapter 10
Process Variation Tolerant Single-supply True Voltage Level Shifter

10.1 Introduction

System-on-chip (SoC) solutions and multi-core computing architectures are becoming increasingly common in many common applications. For such computing paradigms, energy and power minimization is a crucial design goal. Both the dynamic and the leakage power consumption of a CMOS circuit depend upon the supply voltage, and they decrease at least quadratically with decreasing supply voltages. Therefore, in recent times, it is common to decrease the supply voltage value in noncritical parts of SoCs and multi-core processors, to reduce the power and energy consumption. This results in a situation where the many blocks in an SoC design operate at different supply voltage levels, to minimize system power and energy values [1,2]. Similarly, multi-core processors have different cores operating at different supply voltage values, depending on the computational demand.

When a signal traverses on-chip voltage domains, a level shifter is required. Inverters can handle a high to low voltage shift with low delays and minimal leakage. For a low to high voltage level translation, inverters tend to consume a large amount of leakage power, and hence special circuits are needed for this type of level translation. Such special circuits are called voltage level shifters (VLS). Moreover, different blocks/cores in SoCs and multi-core processors may employ dynamic voltage scaling (DVS) to meet the variable speed/power requirements at different times [3,4,5]. As a consequence, many voltage domains are formed on a single IC or SoC, each operating at different supply voltage values at different times of the computation. Therefore, the VLS required to interface these voltage domains should be able to efficiently convert any voltage level to any other desired voltage level, and the voltage of the input to the VLS can in general be either *greater* than or *less* than the voltage of the output. Also, since the key device parameters vary significantly in the DSM era, it is required that the performance of the voltage level shifter does not vary significantly due to these device parameter variations.

The rest of the chapter is organized as follows. The need for a novel single supply, process variation tolerant, voltage level shifter (SS-VLS) is highlighted in Sect. 10.2. Section 10.3 discusses related previous work in the design of an SS-VLS. In Sect. 10.4, the design of the proposed process variation tolerant voltage level

R. Garg and S. P. Khatri, *Analysis and Design of Resilient VLSI Circuits: Mitigating Soft Errors and Process Variations*, DOI 10.1007/978-1-4419-0931-2_10,

shifter is described. The proposed voltage level shifter uses only one supply voltage, and it can convert any voltage level to any other desired voltage level with low delay and power. Hence, it is referred to as a single supply true[1] voltage level shifter (SS-TVLS). In Sect. 10.5, experimental results are presented, which demonstrate that SS-TVLS outperforms the best known previous approach. Finally, a summary of this chapter is presented in Sect. 10.6.

10.2 The Need for a Single-supply Voltage Level Shifter

A conventional voltage level shifter (CVLS) is shown in Fig. 10.1. It requires two voltage supplies: the input domain voltage supply (VDDI) and the output domain voltage supply (VDDO). The operation of circuit is as follows. When the input signal *in* is at the VDDI value (*inb* is at the GND value), MN1 turns ON (MN2 is off). Thus pulls the *outb* signal to GND. This transition of the *outb* signal turns on MP2, which pulls up the *out* signal to the VDDO value. When *in* is at GND (*inb* is at the VDDI value), MN1 is off and MN2 is on, which turns on MP1. MP1 pulls up *outb* to the VDDO value. Although there are no high leakage paths from VDDO to GND in this circuit, both VDDI and VDDO are required for the voltage level conversion. This can be a hard requirement to satisfy, especially if the VDDO and VDDI domains are separated by a large physical distance. Supply voltage wires typically need

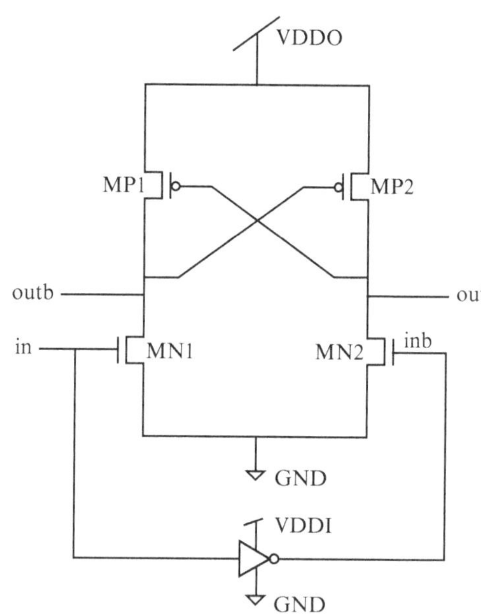

Fig. 10.1 Conventional voltage level shifter

[1] The SS-TVLS is *true* in the sense that it can perform level conversion of a signal from a lower voltage domain to a higher voltage domain and vice versa.

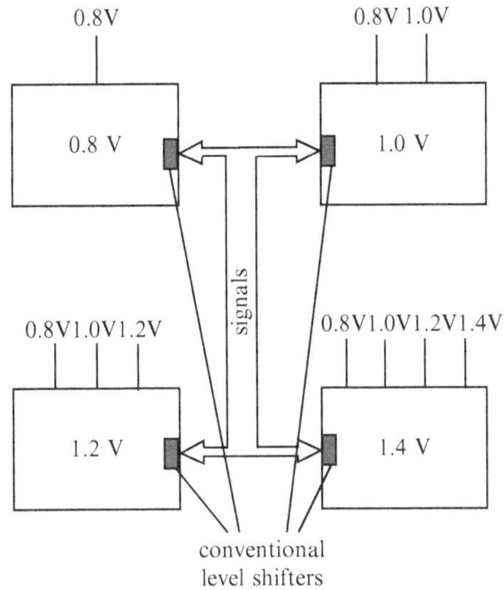

Fig. 10.2 Multivoltage system using CVLS

to be quite wide (especially if VDDO and VDDI are physically far apart), resulting in a large area penalty. Figure 10.2 shows a multivoltage system where four modules are interacting with each other using CVLS. A voltage level conversion at the input of a particular voltage domain (V) will require all the supply voltages of other voltage domains $\{W\}$, which drive at least one signal to V, and whose voltage level is lower than the voltage level of V. This may result in routing congestion, excessive area utilization, and also may pose restrictions on module placement. From the schematic diagram of the CVLS shown in Fig. 10.1, it can be observed that the routing of additional supply voltages can be avoided by transmitting a differential signal to a different voltage domain (i.e., by transmitting both *in* and *inb*). However, this strategy would require one additional wire per signal and hence could lead to routing congestion as well. This problem is further aggravated by the increasing number of voltage domains in SoCs and multi-core architectures. Additional complexity is encountered if the voltage domains have *variable* voltages, which requires a domain to receive the supply voltages of *every other* domain in the system. In such a scenario, it is not known apriori whether VDDI < VDDO or VDDI > VDDO. Therefore, a single supply voltage level shifter (SS-VLS) is desired, to convert any voltage level to any other desired voltage level with a predictable delay and low power, utilizing the supply voltage of the VDDO domain alone. In addition, such a VLS should be "true" (i.e., operate for both VDD < VDDO and VDDI > VDDO). One such solution is proposed in this monograph, and is referred to as a single supply true voltage level shifter (SS-TVLS) previously published [6]. The use of a single supply voltage (VDDO) for level conversion would help ease placement and routing constraints, enabling efficient physical design of the IC. This would also help in reducing the number of input and output pins of a block.

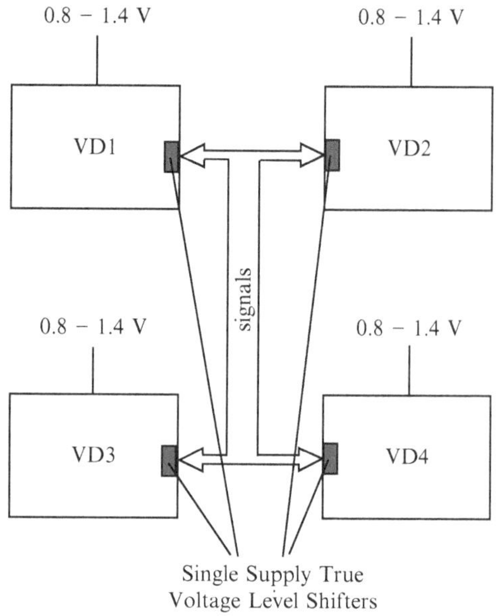

Fig. 10.3 Multivoltage system using SS-TVLS

Figure 10.3 shows a multivoltage system, where four modules (with DVS) interact with each other using SS-TVLS. Note that since the performance of the voltage level shifter is crucial for the performance of the overall system, the voltage level shifter should perform reliably under process variations. In this monograph, the devices of the proposed SS-TVLS were carefully sized to increase its tolerance to process variations, while maintaining a low leakage and low power consumption. This is quantified through experimental results in Sect. 10.5.2.

10.3 Related Previous Work

Several kinds of VLS have been proposed over the years, to minimize power consumption [7, 8, 9, 10]. Most of these approaches utilize dual supply voltages, which make them unattractive for SoCs and multi-core architectures for reasons already discussed. The work of [7] was focused on using bootstrapped gate drivers to minimize voltage swings. This helps in reducing the switching power consumption in the conventional level shifter, and also helps in increasing the speed of the level shifter. In [8], the authors proposed a method of incorporating voltage level conversion into regular CMOS gates by using a second threshold voltage. They proposed a scheme to modify the threshold voltage of high voltage gates (which are driven by the outputs of low voltage gates) to achieve the level shifting functionality along with the logical operation. This work focused on reducing power while using dual supply voltages. In [9], Wang et al. proposed a level up-shifter along with a level down-shifter, to interface 1.0 and 3.3 V voltage domains. The level up-shifters use zero-V_t thick oxide NMOS devices to clamp the voltage to protect the 1 V NMOS switches from high voltage stress across the gate oxide. The level down-shifter used thick oxide NMOS devices with 1 V supplies as both pull-up and pull-down devices. This approach also requires dual supply voltages. In [10], the authors presented a low-to-high voltage level shifter for use in a VLSI chip for MEMS applications. The design uses a stack of devices in series between the rail voltages, biased by five different bias voltages for the conversion.

The SS-VLS proposed in [11] uses a diode connected NMOS device between the supply and output, to convert a low level to a high voltage level. There is a threshold voltage drop in this diode-connected NMOS device, which reduces the supply voltage to the input inverter. This level shifter has a limited range of operation, and suffers from higher leakage currents when the difference in the voltage levels of the output supply and the input signal is larger than the threshold voltage. In [12], the authors present a SS-VLS design, which tries to address the issues associated with the design of [11]. However, their SS-VLS is only able to convert a low voltage domain signal to a higher voltage domain (VDDI < VDDO). Also, the leakage currents of the SS-VLS are relatively high. In contrast to these SS-VLS implementations, the SS-TVLS proposed in this monograph can convert *any* voltage level to any other desired voltage level (i.e., it is a "true" voltage shifter) *without using any control signals*. At the same time, the leakage currents of the proposed SS-TVLS design are very low. Note that none of the previous approaches have analyzed the performance of their VLS under process variations. The performance of the proposed SS-TVLS (under process variations) is compared with the best known previous approach (in terms of functionality) in Sect. 10.5.2.

10.4 Proposed Single-supply True Voltage Level Shifter

The schematic diagram of the proposed SS-TVLS is shown in Fig. 10.4. This SS-TVLS was implemented using a 90 nm PTM [13] technology. Note that devices with thick channel lines are high-V_T devices. Their V_T is 0.49 V for NMOS and −0.44 V for PMOS, while the nominal V_T is 0.39 V for NMOS and −0.34 V for PMOS. Also note that the NOR gate shown in Fig. 10.4 uses the VDDO supply. The sizes (width/length) of all devices (in μm) are shown in the figure. The substrate terminals of all PMOS (NMOS) devices in this figure are connected to VDDO (GND). The operation of the SS-TVLS can be explained by considering two scenarios. The timing diagram of the SS-TVLS is shown in Fig. 10.5, and it is applicable to both scenarios. In the first scenario, VDDO > VDDI (i.e., the VLS has to convert a low voltage level to a high voltage level). In this case, when the input signal *in* goes high (to the VDDI value), the output node *outb* starts falling due to the NOR gate. However, the PMOS transistor of the NOR gate whose gate terminal is driven by *in* is not in complete cut-off (i.e., it is leaking) because VDDI < VDDO. Thus, there is temporary leakage path between VDDO and GND, which is eliminated by the rising of *node*2 (the second input of the NOR gate) to the VDDO value. After the input

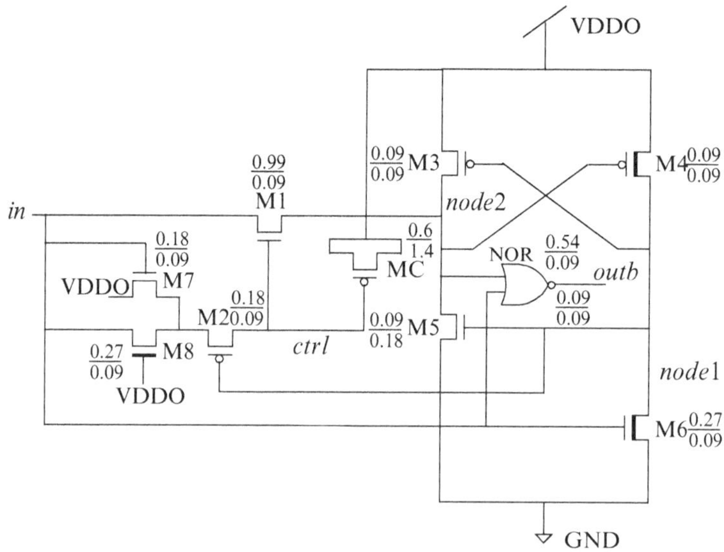

Fig. 10.4 Novel single supply true voltage level shifter

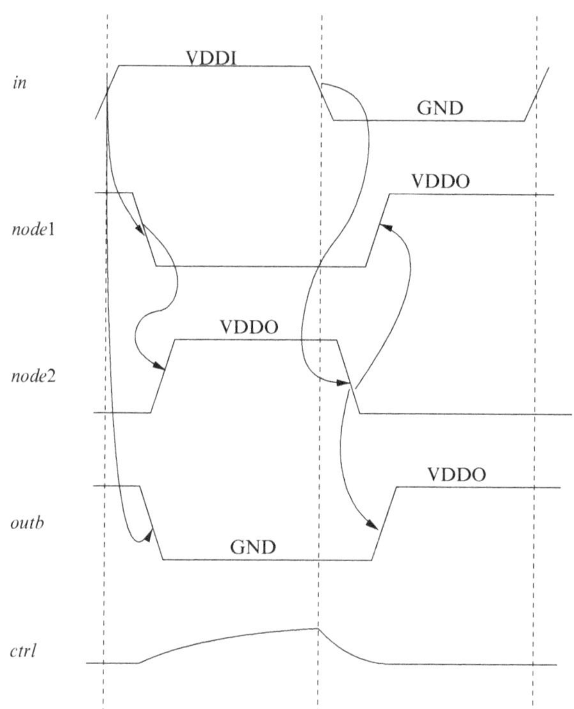

Fig. 10.5 Timing diagram for the proposed SS-TVLS

signal *in* goes high, M6 turns on and thus pulls *node*1 to GND. This causes M3 to turn on and hence *node*2 is pulled up to VDDO, causing the output node *outb* to be pulled down to GND. The previously mentioned leakage path between VDDO and GND is removed when *node*2 is pulled to to VDDO. During this phase, as *in* is high and it is at VDDI (< VDDO), M8 is ON along with M2, which results in the charging of the *ctrl* node (whose capacitance is dominated by the gate capacitance of MC) to a value which is the minimum of VDDI and VDDO-V_T^{M8} (where V_T^{M8} is the threshold voltage of M8). Note that M1, M4, M5, and M7 are turned off when *in* is at the logic high value.

Now when the *in* node falls, M6 turns off while M1 turns on (because the gate to source voltage of M1 is more than V_T^{M1}). This leads to the discharge of *node*2 (and the charging of *node*1) and thus the NOR gate output rises to VDDO (since both the inputs of the NOR gate are at the GND value). In this phase, M3, M2, M6, and M7 are turned off while M4 and M5 are turned on. The *ctrl* node discharges through M2 and M8 during the time when M2 is turning off. The node capacitance of *ctrl* (implemented as the gate capacitance of MC) is selected to be large enough to allow the discharge of *node*2. Note that the NOR gate is used to balance the rising and the falling delays of the SS-TVLS. It also provides the SS-TVLS the same load driving capability as a minimum size inverter. Note that the SS-TVLS is an inverting voltage level shifter. An extra inverter is not required at the output of the SS-TVLS because this polarity inversion can be subsumed in the logic of the VDDO voltage domain. In the experiments, the VLS approach that the SS-TVLS is compared with has the same inverting property.

In the second scenario, the SS-TVLS performs the conversion of a high voltage level to a low voltage level (i.e., VDDO < VDDI). In this scenario as well, when the input *in* goes high to the VDDI value, the output node *outb* falls to the GND value. In this scenario, as VDDI > VDDO, the PMOS transistor of the NOR gate whose gate terminal is driven by *in* is in deep cut-off, and hence there is no leakage path between VDDO and GND. After *in* goes high to VDDI, M6 turns and pulls down its drain node. This turns on M3 which then charges *node*2 to VDDO. During this phase, as VDDI > VDDO therefore, M7 is ON and M2 is also ON. M8 is off in this case. Thus, the *ctrl* node voltage charges to a value *min*(VDDO, VDDI-V_T^{M7}). Here V_T^{M7} is the threshold voltage of M7. Note that M1, M4, and M5 are turned off when *in* is at VDDI. The rest of the operation of the SS-TVLS when *in* transitions to GND is identical to the first scenario. Note that the SS-TVLS works for VDDI > VDDO as well as VDDI < VDDO because M1 never turns on when *in* is logically high (regardless of whether VDDI > VDDO or VDDI < VDDO).

The SS-TVLS exhibits very low leakage currents when compared with the best known voltage level shifter [12] for VDDI < VDDO. There are several reasons for this. Note that the devices M4 and M6 are high V_T devices, to reduce leakage currents. Also, all the devices of the proposed SS-TVLS were carefully sized to improve its tolerance to process variations and further, the tradeoff between speed and leakage power was considered. Specifically, to improve the variation tolerance of the SS-TVLS, the transistors M1, M3, M6, and MC were carefully sized since they are very critical for the performance of SS-TVLS. As mentioned earlier, the maximum voltage value that the *ctrl* node can charge to is the minimum of VDDI and VDDO-V_T^{M8} when VDDI < VDDO, and VDDO and VDDI-V_T^{M7} when VDDI > VDDO, respectively. Thus, when the voltage values of the VDDI and VDDO domains are small and close to each other, then the *ctrl* node charges to VDDO-V_T^{M8}. Therefore, a low V_T NMOS device[2] is used for M8 to ensure that *ctrl* can charge to a sufficiently large voltage value. This also helps in increasing the voltage translation range of the SS-TVLS. Note that all other transistors (M1, M2, M3, M5, MC, and M7 and NOR gate transistors) are nominal V_T devices.

10.5 Experimental Results

The proposed SS-TVLS was simulated using SPICE [14], with a 90 nm PTM [13] model card. An inverter is the best level shifter when VDDI > VDDO. However, if VDDI < VDDO, the inverter cannot be used, because of the high leakage currents that result in such a conversion. For such a scenario, the best known previous approach [12] yields low leakage currents. Therefore, to compare the performance of SS-TVLS, a combination of an inverter and the SS-VLS of [12] (as shown in Fig. 10.6) was simulated. For the SS-VLS of [12], the devices used are of the same size as reported in [12]. Note that the combined VLS of Fig. 10.6 requires a control signal, which indicates whether VDDI is greater or smaller than VDDO. The SS-TVLS does not require such a signal. For all simulation results, both the SS-TVLS and the combined VLS were driven by identical inverters.

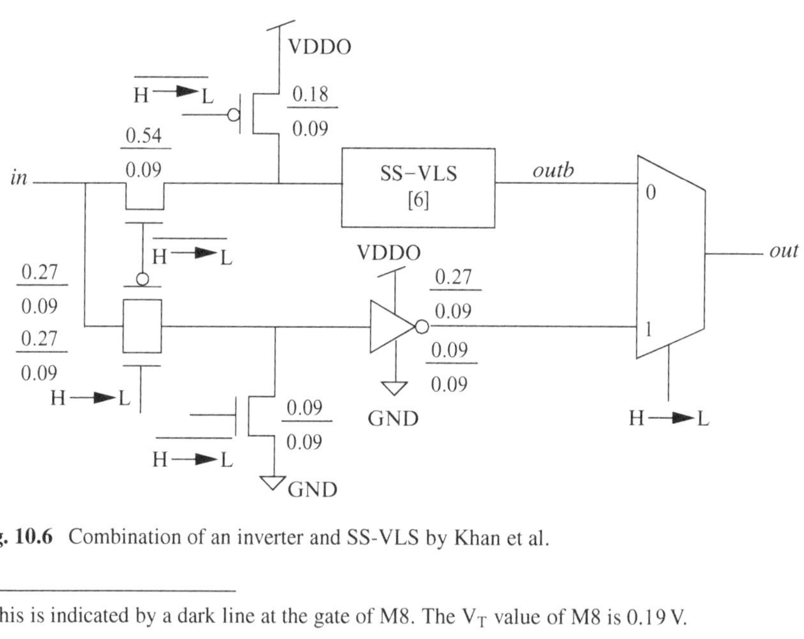

Fig. 10.6 Combination of an inverter and SS-VLS by Khan et al.

[2] This is indicated by a dark line at the gate of M8. The V_T value of M8 is 0.19 V.

Note that the delays of the SS-TVLS as well as the SS-VLS of [12] are dependent on the input sequence. The worst-case is a 0-1-0-1-0... sequence on the inputs. For this sequence, the voltage achieved at the *ctrl* node when the input switches to 0 is the lowest across all sequences, resulting in a higher output rising delay. The delay numbers reported in this chapter are the worst-case delays across all possible input sequences.

10.5.1 Performance Comparison with Nominal Parameters Value

Table 10.1 reports the results obtained for voltage level shifting from 0.8 to 1.2 V at a temperature of 27°C. Column 1 reports the performance parameter under consideration. Column 2 reports the results obtained for the proposed SS-TVLS. Column 3 reports the results obtained for the combined VLS of Fig. 10.6. Column 4 reports the ratio of the results obtained for the combined VLS compared with the corresponding results for the SS-TVLS. Note that the rising (falling) delay is defined as the delay of the rising (falling) output signal. Similarly, "Leakage current high (low)" in the table represents the leakage current when the output signal is at VDDO (GND) value. From Table 10.1, we observe that the SS-TVLS performs significantly better than the combined VLS in terms of delay (5.6× faster for a rising output and 1.5× faster for a falling output), power (2.6× lower for a rising output, and 3.5× lower for a falling output) and leakage (7.6× lower for a high output, and 19.8× lower for a low output).

Table 10.2 reports the results obtained for voltage level conversion from 1.2 to 0.8 V at a temperature of 27°C. Column 1 reports the performance parameter under consideration. Column 2 reports the results obtained for the proposed SS-TVLS. Column 3 reports the results obtained for the combined VLS shown in Fig. 10.6. Column 4 reports the ratio of the results obtained for the combined VLS compared with the corresponding results for the SS-TVLS. As reported in Table 10.2, the proposed SS-TVLS performs very well when compared with the combined VLS of Fig. 10.6 with very low leakage currents (4.5× lower for a high output, and 9.3× lower for a low output). Also it is faster than the combined VLS

Table 10.1 Low to high level shifting

Performance parameter	Proposed SS-TVLS	Combined VLS of Fig. 10.6	Ratio (combined VLS/SS-TVLS)
Delay rise (ps)	22.0	122.6	5.6
delay fall (ps)	33.3	50.5	1.5
Power rise (μW)	27.6	71.87	2.6
Power fall (μW)	33.8	119.27	3.5
Leakage current high (nA)	20.8	157.2	7.6
Leakage current low (nA)	3.6	71.1	19.8

Table 10.2 High to low level shifting

Performance parameter	Proposed SS-TVLS	Combined VLS of Fig. 10.6	Ratio (combined VLS/SS-TVLS)
Rise delay (ps)	34.9	46.5	1.3
Fall delay (ps)	15.7	35.2	2.2
Power rise (μW)	27.3	20.7	0.8
Power fall (μW)	59.3	56.8	1.0
Leakage current high (nA)	7.3	32.5	4.5
Leakage current low (nA)	3.9	36.3	9.3

(1.3× faster for a rising output and 2.2× faster for a falling output). Note that the delay of the combined VLS is the summation of the delays of the transmission gate (at the input side), the multiplexer (at the output side) and the inverter. Therefore, the delay of the combined VLS is more than the inverter delay alone.

10.5.2 Performance Comparison Under Process and Temperature Variations

To demonstrate the process variation tolerance of the SS-TVLS, the SS-TVLS was simulated under process and temperature variations. The temperature, the channel width, the channel length, and the threshold voltage of all devices in the SS-TVLS were varied. The temperature of all the devices were varied together, while all other parameters were varied independently. For channel lengths and widths, the mean was taken to be equal to the nominal value and the standard deviation used was taken to be 3.34% of the L_{min} of the process (i.e., 90 nm). For threshold voltage, the mean was taken to be equal to the nominal value and the standard deviation used was taken to be 3.34% of the nominal value (so that the three times of the standard deviation is 10% of the nominal value). Three different values of temperature were used (27°C, 60°C, and 90°C). One thousand Monte-Carlo simulations were performed for both cases, i.e., for high-to-low and low-to-high voltage conversion. These simulations were performed at each of the three temperatures mentioned above. In all Monte-Carlo simulation, the SS-TVLS was able to convert the voltage level correctly for all samples. The outputs of both designs were loaded with a fixed capacitance of 1 fF.

The results obtained from the 1,000 Monte-Carlo simulations conducted at a temperature of 27°C are reported in Table 10.3, for low-to-high and high-to-low voltage level conversion. In Table 10.3, Column 1 reports the performance parameter under consideration. Columns 2–5 report the results for low-to-high voltage level conversion and Columns 6–9 report the results for high-to-low voltage level conversion. Columns 2 and 3 report the mean and the standard deviation of the values obtained for the proposed SS-TVLS. Columns 4 and 5 report the mean and the standard deviation for the combined VLS shown in Fig. 10.6. Columns 6–9 report the same results as Columns 2–5 but for high-to-low voltage level conversion. From

Table 10.3 Process variations simulation results for low-to-high and high-to-low level shifting at $T = 27°C$

	Low to high				High to low			
	Proposed SS-TVLS		Combined VLS of Fig. 10.6		Proposed SS-TVLS		Combined VLS of Fig. 10.6	
Performance parameter	μ	σ	μ	σ	μ	σ	μ	σ
Delay rise (ps)	22.08	1.1	129.4	27.4	35.1	2.4	52.0	3.9
Delay fall (ps)	33.2	1.9	50.4	6.0	15.6	0.8	34.8	1.3
Power rise (μW)	27.7	0.8	78.9	7.3	27.5	1.3	22.5	1.1
Power fall (μW)	33.8	0.4	114.2	7.2	59.5	0.6	52.5	0.9
Leakage current high (nA)	31.5	13.7	218.8	158.6	8.6	3.0	41.4	14.1
Leakage current low (nA)	3.8	3.8	102.9	75.4	3.6	1.3	32.3	9.0

Table 10.4 Process variations simulation results for low-to-high and high-to-low level shifting at $T = 60°C$

	Low to high				High to low			
	Proposed SS-TVLS		Combined VLS of Fig. 10.6		Proposed SS-TVLS		Combined VLS of Fig. 10.6	
Performance parameter	μ	σ	μ	σ	μ	σ	μ	σ
Delay rise (ps)	18.5	0.8	131.3	39.4	29.1	2.0	43.2	2.9
Delay fall (ps)	29.9	1.5	48.4	5.7	14.3	0.6	30.3	1.0
Power rise (μW)	27.4	0.7	86.0	11.1	27.1	1.1	22.4	1.0
Power fall (μW)	33.6	0.34	123.7.2	8.3	60.1	0.7	53.3	0.9
Leakage current high (nA)	30.4	13.4	202.4	130.2	7.9	3.2	40.3	12.8
Leakage current low (nA)	3.7	3.8	98.5	61.1	3.2	1.5	32.9	8.4

this table, observe that the mean delay and power are closer to their nominal values. However, the mean value of the leakage current is different from the nominal value, which is due to an exponential dependence of the leakage current with the threshold voltage. The standard deviation of all performance parameters, i.e., delay, power, and leakage current is much lower for the SS-TVLS when compared with the combined VLS of Fig. 10.6. This demonstrates that the SS-TVLS is more tolerant to process and temperate variations than the combined VLS. Monte-Carlo simulation results for other temperatures are presented in Tables 10.4 (for $T = 60°C$) and 10.5 (for $T = 90°C$). Note that the results are substantially similar to those obtained at $T = 27°C$ (Table 10.3).

10.5.3 Voltage Translation Range for SS-TVLS

To evaluate the effectiveness of the SS-TVLS for SoCs and multi-core processors having multiple voltage domains with DVS, VDDI, and VDDO values were varied from 0.8 to 1.4 V in steps of 5 mV. The SS-TVLS was simulated for all VDDI and VDDO combinations. The SS-TVLS was able to translate voltage levels efficiently for all VDDI and VDDO combinations. Figures 10.7 and 10.8 show the plot of rising

Table 10.5 Process variations simulation results for low-to-high and high-to-low level shifting at $T = 90°C$

	Low to high				High to low			
	Proposed SS-TVLS		Combined VLS of Fig. 10.6		Proposed SS-TVLS		Combined VLS of Fig. 10.6	
Performance parameter	μ	σ	μ	σ	μ	σ	μ	σ
Delay rise (ps)	16.3	0.6	146.7	54.2	26.4	1.9	36.9	2.3
Delay fall (ps)	27.8	1.3	47.8	5.9	13.5	0.6	27.4	0.9
Power rise (μW)	27.3	0.6	96.8	16.1	23.4	0.8	22.4	0.8
Power fall (μW)	33.6	0.35	134.8	9.4	51.4	0.6	53.9	1.1
Leakage current high (nA)	28.1	11.3	200.8	128.6	7.6	3.1	39.7	13.9
Leakage current low (nA)	3.4	1.9	94.0	66.3	3.1	1.3	33.3	8.6

and falling delays and powers when VDDI and VDDO were varied between 0.8 and 1.4 V. The leakage current for a high output value and a low output value is shown in Fig. 10.9 for VDDI and VDDO varying between 0.8 and 1.4 V. These figures show that the rising and the falling delays and powers change smoothly with changing VDDI and VDDO values, over the entire voltage range. Similarly, the leakage current for a high output and a low output is also well behaved across the operating range as shown in Fig. 10.9. Therefore, it can be concluded that the SS-TVLS can effectively perform voltage level translation over a wide range of VDDI and VDDO voltage values, and hence it is suitable for SoCs and multi-core processors.

As mentioned earlier, the maximum voltage value that the *ctrl* node can charge to is the minimum of VDDI and VDDO-V_T^{M8} when VDDI < VDDO, and VDDO and VDDI-V_T^{M7} when VDDI > VDDO. Therefore, none of the diffusion-bulk diodes of any device (both PMOS and NMOS transistors of Fig. 10.4) get forward biased for any values of VDDI and VDDO. Thus, the voltage translation range is not limited by the diode turn on voltage.

It is possible to increase the voltage translation range, i.e., from a value less than 0.8–1.4 V. To achieve this, the size of M6 needs to be increased. However, when VDDI or VDDO < 0.8 V, then the maximum voltage reached at the *ctrl* node of SS-TVLS of Fig. 10.4 is small, and hence M1 is not turned on sufficiently to discharge *node*2 when *in* falls to GND. To address this, the bulk terminals of M7 and M8 transistors can be connected to their respective source terminals instead of GND. This will help in avoiding the body effect seen by M7 and M8 and will increase the maximum voltage value achieved at the *ctrl*. Note that in the SS-TVLS shown in Fig. 10.4, the body terminals of M7 and M8 are connected to GND.

10.5.4 Layout of SS-TVLS

The layout of the proposed SS-TVLS was created in the Cadence Virtuoso layout editor and is shown in Fig. 10.10. A layout versus schematic (LVS) check was done.

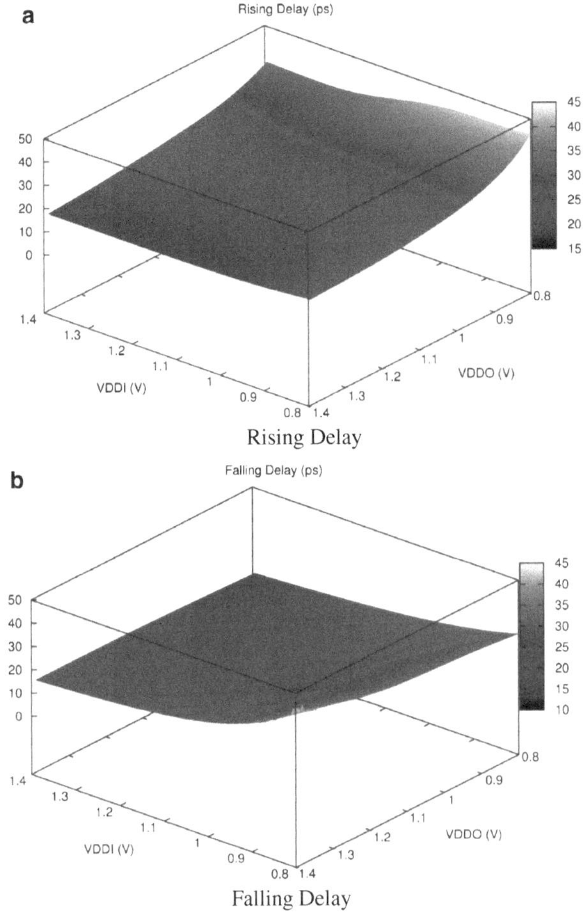

Fig. 10.7 Delay of SS-TVLS: (**a**) rising, (**b**) falling

The layout area of SS-TVLS is $4.47\mu m^2$ (the width is $0.837\mu m$ and the height is $5.355\mu m$), which is lower that the layout area of the combined VLS ($\sim 12.53\mu m^2$). The sizes of all the devices of the SS-TVLS are shown in Fig. 10.4. Note that the devices of the SS-TVLS were sized considering the tradeoff between delay and leakage power, while maximizing the process and temperature variation tolerance of the SS-TVLS.

The experimental results clearly demonstrate that the proposed SS-TVLS performs much better than the combined VLS of Fig. 10.6. When it is not known apriori whether VDDI < VDDO or VDDI > VDDO, then the SS-TVLS offers a great advantage over the combined VLS of Fig. 10.6, because of its significantly lower leakage currents (7.6× (4.5×) lower for a high output, and 19.8× (9.3×) lower for a low output, when VDDI < VDDO (VDDI > VDDO)). Moreover, the SS-TVLS

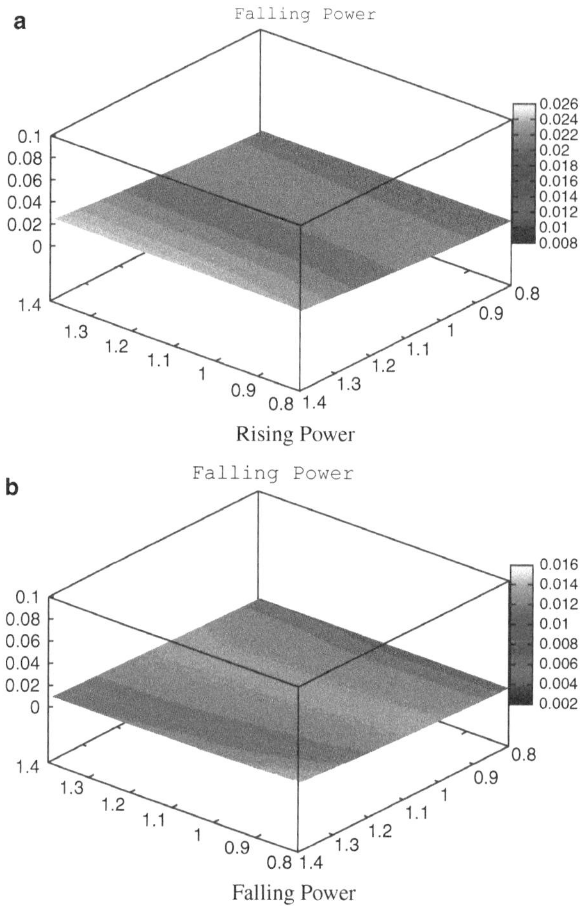

Fig. 10.8 Power of SS-TVLS: (**a**) rising, (**b**) falling

does not require any control signals and only requires the VDDO supply. This helps in reducing the circuit complexity and also helps in placement and routing.

10.6 Chapter Summary

Modern ICs often have several voltage domains. Whenever a signal traverses voltage domains, a level shifter is required. Moreover, these ICs often employ dynamic voltage scaling, because of which it may not be possible to know apriori if a high-to-low or low-to-high voltage level conversion is required. Thus, an efficient voltage level shifter is required, which can convert any voltage level to any other desired

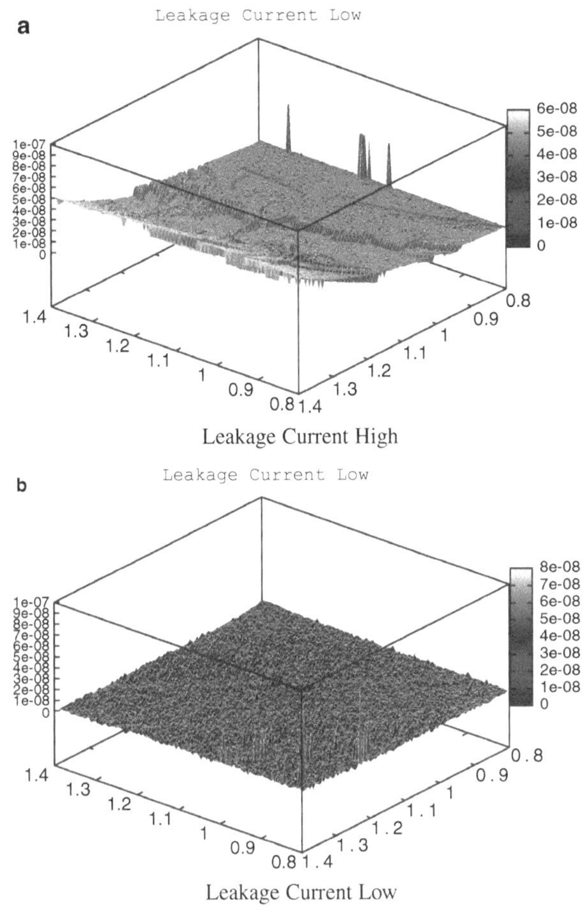

Fig. 10.9 Leakage current of SS-TVLS: (**a**) high, (**b**) low

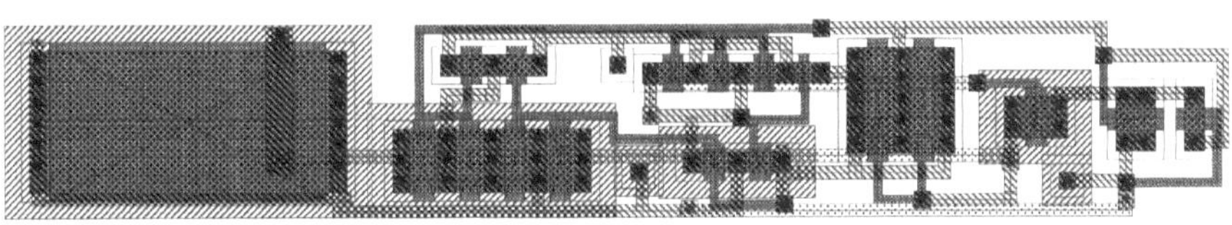

Fig. 10.10 Layout of the proposed SS-TVLS

voltage level. Also, since process variations are increasing with device scaling, the voltage level shifter should be tolerant to process and temperature variations.

In this chapter, a process and temperature variation tolerant single-supply "true" voltage level shifter (SS-TVLS) was presented, which can handle both low-to-high and high-to-low voltage translations. The use of a single supply voltage reduces layout congestion by eliminating the need for routing both supply voltages.

The proposed circuit was simulated in a 90 nm technology using SPICE. Simulation results demonstrate that the proposed SS-TVLS performs much better than the combined VLS of Fig. 10.6. The combined VLS uses an inverter for high-to-low voltage translation and the best known previous approach [12] for low-to-high voltage level shifting. Also, the proposed SS-TVLS is more tolerant to process and temperature variations than the combined VLS.

References

1. D. Lackey, David E. Lackey, Paul S. Zuchowski, Thomas R. Bednar, Douglas W. Stout, Scott W. Gould, and John M. Cohn, "Managing power and performance for SOC designs using voltage islands," in *Proc. of the Intl. Conf. on Computer-Aided Design*, Nov. 2002, pp. 195–202.
2. T. Hattori et. al., "A power management scheme controlling 20 power domains for a single-chip mobile processor," in *Proc. of the Intl. Solid State Circuits Conf.*, Feb. 2006, pp. 540–541.
3. W. Kim, D. Shin, H. S. Yun, J. Kim, and S. L. Min, "Performance comparison of dynamic voltage scaling algorithms for hard real-time systems," in *Proc. of the Symposium on Real-Time and Embedded Technology and Applications*, 2002, pp. 219–228.
4. B. Zhai, D. Blaauw, D. Sylvester, and K. Flautner, "Theoretical and practical limits of dynamic voltage scaling," in *Proc. of the Design Automation Conf.*, 2004, pp. 868–873.
5. C. Duan and S. P. Khatri, "Computing during supply voltage switching in DVS enabled real-time processors," in *Proc. of the Intl. Symposium on Circuits and Systems*, May 2006, pp. 5115–5118.
6. R. Garg, Gagandeep Mallarapu, and S. P. Khatri, "A single-supply true voltage level shifter," in *Proc. of the Conf. on Design Automation and Test in Europe*, March 2008, pp. 979–984.
7. S. C. Tan and X. W.Sun, "Low power CMOS level shifters by bootstrapping technique," *IEEE Electronics Letters*, pp. 876– 878, Aug. 2002.
8. A. U. Diril, Y. S. Dhillon, A. Chatterjee, and A. D. Singh, "Level-shifter free design of low power dual supply voltage CMOS circuits using dual threshold voltages," *IEEE Transaction on Very Large Scale Integration*, vol. 13, pp. 1103–1107, Sept. 2005.
9. W. Wang, M. Ker, M. Chiang, and C. Chen, "Level shifters for high-speed 1-V to 3.3-V interfaces in a 0.13-pm Cu-interconnectiod/low-k CMOS technology," in *Proc. of the Intl. Symposium on VLSI Technology, Systems, and Applications*, 18–20 April 2001, pp. 307–310.
10. D. Pan, H. W. Li, and B. M. Wilamowski, "A low voltage to high voltage level shifter circuit for MEMS application," in *Proc. of the Biennial University/Government/Industry Microelectronics Symposium*, July 2003, pp. 128–131.
11. R. Puri, L. Stok, J. Cohn, D. S. Kung, D. Z. Pan, D. Sylvester, A. Srivastava, and S. Kulkarni, "Pushing ASIC performance in a power envelope," in *Proc. of the Design Automation Conf.*, June 2003, pp. 788– 793.
12. Q. A. Khan, S. K. Wadhwa, and K. Misri, "A single supply level shifter for multi voltage systems," in *Proc. of the Intl. Conf. on VLSI Design*, Jan. 2006, pp. 1–4.
13. Nanoscale integration and modeling (NIMO) group (2007), *ASU Predictive Technology Model [On-line]*, Available: http://www.eas.asu.edu/~ptm.
14. L. Nagel, "Spice: A computer program to simulate computer circuits," in *University of California, Berkeley UCB/ERL Memo M520*, May 1995.

Chapter 11
Conclusions and Future Directions

The focus of this monograph is to improve the reliability of VLSI circuits by addressing two major issues (radiation particle strikes and process variations) encountered in the deep submicron era. This monograph developed several analysis and design approaches to facilitate the realization of VLSI circuits, which are tolerant to the radiation particle strikes and process variations. This chapter summarizes the work presented in this monograph and presents some avenues for further research.

Chapter 1 described the effects of the radiation particle strikes and process variations on VLSI circuits, thereby motivating the need to address these issues. This chapter also provided the background information about these two topics, and also highlighted their relevance in future technologies. This monograph consists of two parts. The first part (Chaps. 2–7) of the monograph presented four analysis and two design approaches to address the radiation issue. The second part (Chaps. 8–10) addressed the process variation issue by presenting one analysis and two design approaches. Thus, several analysis and design techniques are presented in this monograph, significantly augmenting the existing work in the area of resilient VLSI circuit design.

Two analysis approaches were developed in this monograph to analyze the effects of radiation particle strikes in combinational circuits (Chaps. 2 and 3). Chapter 2 presented a model to estimate the pulse width of the radiation-induced voltage glitch, and Chap. 3 described a model to approximate the shape of the radiation-induced voltage glitch. Both these approaches used more accurate gate models (than a linear RC gate model), and also considered the ion track establishment constant (τ_β) of the radiation-induced current pulse in the analysis. The previous modeling approaches [1, 2] used a linear RC gate model, and ignored τ_β which increases the inaccuracy of their analysis. Therefore, the modeling approaches presented in Chaps. 2 and 3 are more accurate than the previous approaches. The proposed models are fast and accurate, and thus can easily be incorporated in a design flow to implement radiation tolerant circuits.

It was mentioned in Chap. 3 that there exist efficient tools which can propagate voltage glitches in a design. However, it may be possible to implement more efficient tools exclusively for propagating radiation-induced voltage glitches. Therefore, a possible future direction could be to develop efficient radiation-induced

R. Garg and S. P. Khatri, *Analysis and Design of Resilient VLSI Circuits: Mitigating Soft Errors and Process Variations*, DOI 10.1007/978-1-4419-0931-2_11,

voltage glitch propagation tools. Also, the work presented in Chap. 3 only considers radiation particle strikes at the output of a gate. Therefore, this work can be extended to obtain the radiation-induced voltage transients at the output of a gate because of radiation strikes at the internal nodes of the gates. A combination of the models presented in Chaps. 2 and 3, along with a voltage glitch propagation tool, and an approach that estimates the radiation-induced transients at the output of a gate due to radiation strikes at the internal nodes, can be used for hardening a circuit efficiently. It could also be used for estimating the soft error rate (SER) of a circuit operating in an environment where radiation is present.

SRAM yield is very important from an economic viewpoint, because of the extensive use of memory in modern processors and SoCs. Therefore, SRAM stability analysis tools have become essential. SRAM stability analysis based on static noise margin (SNM) computation often results in pessimistic designs because SNM cannot capture the transient behavior of the noise. Therefore, to improve accuracy, dynamic stability analysis techniques are required. A model was developed in this monograph to perform the dynamic stability analysis of an SRAM cell in the presence of a radiation particle strike, as described in Chap. 4. This model utilizes a double exponential current equation for modeling a radiation particle strike, and it is able to predict (more accurately than [3]) whether a radiation particle strike will result in a state flip in a 6T-SRAM cell (for given values of Q, τ_α and τ_β). Experimental results demonstrate that this model is very accurate, with a critical charge estimation error of 4.6% compared with HSPICE. The runtime of this model is also significantly lower (by $\sim$2,000$\times$) than the HSPICE runtime. Thus, this model enables an SRAM designer to quickly and accurately analyze the stability of their 6T cell during the design phase.

The model for the dynamic stability of an SRAM cell presented in Chap. 4 considers noise in SRAMs only due to radiation particle strikes. However, there are other types of noise such as power and ground noise, capacitive coupling noise, etc. Therefore, the models similar to the one presented in Chap. 4 are required to perform dynamic stability of an SRAM cell in the presence of capacitive coupling noise, and power and ground noise. In future, the proposed approach for modeling the dynamic stability of an SRAM cell can be extended to include the effects of these noise sources as well. Also, another possible future direction could be to extend the model presented in Chap. 4 to incorporate the effect of process variations on the dynamic stability of SRAMs in the presence of a radiation particle strike.

In recent times, dynamic supply voltage scaling (DVS) has been extensively employed to minimize the power and energy of VLSI systems. Also, subthreshold circuits are becoming more popular. Therefore, the reliability of voltage scaled VLSI systems (when subjected to a radiation event) has become a major concern. With the increasing demand for reliable low power systems, it is necessary to harden DVS and subthreshold circuits efficiently. This makes it necessary to understand the effects of voltage scaling on the radiation tolerance of VLSI systems. To address this, the effects of voltage scaling on the radiation tolerance of VLSI systems was analyzed in Chap. 5. For this analysis, 3D simulations of radiation particle strikes on the output of an inverter (implemented using DVS and subthreshold design) were performed. The radiation particle strike on an inverter was simulated using Sentaurus-DEVICE [4], for different inverter sizes, inverter loads, supply voltages, and radiation particle energies. From these 3D simulations, several nonintuitive observations were made, which are important to consider during the radiation hardening of such DVS and subthreshold circuits. On the basis of these observations, several guidelines were proposed for radiation hardening of such designs. These guidelines suggest that traditional radiation hardening approaches need to be revisited for DVS and subthreshold designs. A charge collection model for DVS circuits was also proposed, which can be used to improve the accuracy of SPICE-based simulations of radiation events in DVS circuits. Note that this work was done for bulk CMOS process.

An extension of this work could be to perform a similar study for 3D devices such as FINFETs. Since the structure and operation of 3D devices is very different from bulk CMOS devices, the effect of voltage scaling on the radiation susceptibility of circuits implemented using these 3D devices might be different compared with bulk CMOS devices.

As mentioned in Chap. 5, a significant amount of the charge gets collected through the diffusion process in DSM devices (since the substrate is heavily doped). Therefore, another possible extension of this work is to perform 3D device simulations to study the effect of different device implementation structures (e.g., instead of implementing one big device, two small devices connected in parallel may be used) on the charge collected due to a radiation particle strike. This study can be useful in hardening a circuit by presenting layout guidelines to enhance radiation resilience.

The results of the analysis of the effects of a radiation particle strike on a circuit can be used for selective hardening of the gates in a circuit, to achieve a desired level of radiation tolerance while satisfying area, delay, and power constraints. For this, efficient circuit level hardening techniques are required. Two hardening approaches were developed in this monograph for combinational circuits, as described in Chaps. 6 and 7. The first hardening approach (referred to as *the diode clamping based approach*) is suitable for hardening circuits against low energy radiation particle strikes, while the second approach (*the split-output based hardening approach*) can harden circuits against very high energy particle strikes. Both these hardening approaches use special gate structures to prevent the occurrence/propagation of the radiation-induced voltage glitch. Also, to keep the area and delay overheads low, only sensitive gates in a combinational circuit were hardened. The gates that were hardened against radiation particle strikes are the gates that contribute significantly to the soft error failure of the circuit. Experimental results presented in Chaps. 6 and 7 demonstrate the effectiveness of these approaches in implementing radiation-tolerant combinational circuits.

In the second part of this monograph, Chap. 8 presented the sensitizable statistical timing analysis methodology developed in this monograph (StatSense) to improve the accuracy of statistical timing analysis of combinational circuits. StatSense improves the accuracy of statistical timing analysis by eliminating false paths in a circuit, and by also using different delay distributions for different input transitions for any gate. Experimental results show that on average, the worst case ($\mu + 3\sigma$) circuit delay reported by StatSense is about 19% lower than that reported by SSTA. Thus, StatSense reduces the pessimism involved in the statistical timing analysis.

The StatSense approach uses Monte Carlo simulations to estimate the delay distribution of a circuit. Therefore, the StatSense approach, although more accurate, is slower than the block-based SSTA approaches. It may be possible to combine the best of the StatSense and the block-based SSTA approaches, to develop a StatSense-like fast block-based SSTA approach. This will help in improving the accuracy of SSTA tools with smaller runtimes.

Process variation tolerant circuit design approaches are required to improve yield and lower manufacturing costs. In Chap. 9, a process variation tolerant design approach for combinational circuits was presented. This approach exploits the fact that random variations can cause a significant mismatch in two identical devices placed next to each other on the die. In this approach, a large gate is implemented by connecting an appropriate number (> 1) of smaller gates in parallel. This parallel connection of smaller gates is referred to as a parallel gate. Since L and V_T variations are largely random and have independent variations in the smaller gates, the variation tolerance of the parallel gate is improved. The parallel gates were implemented as single layout cells. By sharing the diffusion region in the layout of the parallel gates, it is possible to reduce the input and output capacitance of the gates. This helps in improving the nominal circuit delay as well. To keep the area overhead low, only critical gates in a circuit were replaced by their parallel counterparts, to improve the variation tolerance of the circuit. Experimental results from Monte Carlo simulations demonstrate that this process variation tolerant design approach achieves significant improvements in circuit level variation tolerance.

With the increasing usage of dynamic voltage scaling (DVS) in SoCs and multi-core ICs, the number of voltage domains in a single IC or SoC has significantly increased. To interface these voltage domains, voltage level shifters (VLSs) are required. These VLSs should be able to convert any voltage level to any other desired voltage level with a *predictable delay*. Thus, process variation tolerant voltage level shifters are desired. A novel process variation tolerant single-supply true voltage level shifter (SS-TVLS) design was presented in Chap. 10. The SS-TVLS is the first VLS design, which can handle both low-to-high and high-to-low voltage translation without a need for a control signal. The use of a single supply voltage reduces circuit complexity, by eliminating the need for routing 2 supply voltages. The proposed circuit was extensively simulated in a 90 nm technology using SPICE. Simulation results demonstrate that the level shifter is able to perform voltage level shifting with low leakage for both low-to-high, as well as high-to-low voltage level translation. The proposed SS-TVLS is also more tolerant to process and temperature variations compared with a combination of an inverter along with the VLS solution [5]. Thus, the proposed SS-TVLS is better than the best known previous design of VLS approach.

References

1. K. Mohanram, "Closed-form simulation and robustness models for SEU-tolerant design," in *Proc. of the VLSI Test Symposium*, 2005, pp. 327–333.
2. P. Dahlgren and P. Liden, "A switch-level algorithm for simulation of transients in combinational logic," in *Proc. of the Intl. Symposium on Fault-Tolerant Computing*, June 1995, pp. 207–216.
3. B. Zhang, A. Arapostathis, S. Nassif, and M. Orshansky, "Analytical modeling of SRAM dynamic stability," in *Proc. of the Intl. Conf. on Computer-Aided Design*, Nov. 2006, pp. 315–322.
4. Synopsys Inc., Mountain View, CA, *Sentaurus user's manuals*, 2007.12 edition.
5. Q. A. Khan, S. K. Wadhwa, and K. Misri, "A single supply level shifter for multi voltage systems," in *Proc. of the Intl. Conf. on VLSI Design*, Jan. 2006, pp. 1–4.

Appendix A
Sentaurus Related Code

The code used in the work described in Chap. 5 is presented in this appendix. The code used for constructing the 3D NMOS transistor provides detailed information about the physical properties of the NMOS transistor.

A.1 Code for 3D NMOS Device Creation Using Sentaurus-Structure Editor Tool

The code that was used to construct the 3D NMOS transistor using Sentaurus-Structure editor tool is as follows.

```
    ; Setting parameters
; - lateral
(define Lg 0.036) ; [um] Gate length
(define subzmin -4.88); [um] Max. frontside extension in the z-direction
(define subzmax 5.12); [um] Max. backside extension in the z-direction
(define subxmin -5.42); [um] Max. leftside extension in the x-direction
(define subxmax 5.253); [um] Max. rightside extension in the x-direction
(define wn 1); [um] width of the nmos device
(define Lpreox 0.002) ; Poly rexox thickness
(define Lspacer 0.03) ; Spacer length

; Layers
(define Ysub 4) ; [um] Substrate thickness
(define Tox 12e-4) ; [um] Gate oxide thickness
(define Ypol -0.12) ; [um] Poly gate thickness

; Substrate doping level
(define Dop 1e16) ; [1/cm3]
```

```
; Derived quantities
(define Xg (/ Lg 2.0))
(define Ygox (* Tox -1.0))

;———————————————————————————————————
; Overlap resolution: New replaces Old
(isegeo:set-default-boolean "ABA")
;———————————————————————————————————

; CREATE REGIONS
; SUBSTRATE REGION
(isegeo:create-cuboid (position subxmin 0 subzmin) (position subxmax Ysub subz-
max) "Silicon" "region_1" )

; GATE OXIDE REGION - Main
(isegeo:create-cuboid (position (* Xg -1.0) 0 0) (position Xg Ygox wn) "SiO2"
"region_2")

; PolySi GATE - Main
(isegeo:create-cuboid (position (* Xg -1.0) Ygox 0) (position Xg Ypol wn) "PolySi"
"region_3")

; STI REGION - I ("behind" S/D, till the left edge of the gate extension)
(isegeo:create-cuboid (position subxmin 0 wn) (position 0.18 0.4 subzmax) "Oxide"
"STI1" )

; STI REGION - III ("to the right" of S/D)
(isegeo:create-cuboid (position 0.18 0 subzmin) (position 0.93 0.40 subzmax) "Ox-
ide" "STI3" )

; STI REGION - IV ("in front of" of S/D, till the left edge of the gate extension)
(isegeo:create-cuboid (position subxmin 0 0) (position 0.28 0.40 subzmin) "Oxide"
"STI4" )

; STI REGION - VI ("to the left of" of S/D)
(isegeo:create-cuboid (position subxmin 0 0) (position -0.18 0.40 wn) "Oxide"
"STI6" )

; STI REGION - VII ("to the right of p-well contact" of S/D)
(isegeo:create-cuboid (position 1.06 0 subzmin) (position subxmax 0.4 subzmax)
"Oxide" "STI7" )

;Poly Reoxidation
(isegeo:set-default-boolean "BAB")
(isegeo:create-cuboid (position (* (+ Xg Lpreox) -1.0) Ygox 0) (position (+ Xg
```

```
Lpreox) Ypol wn) "Oxide" "PolyReOxide1")
(isegeo:create-cuboid (position (* (+ Xg (+ Lspacer Lpreox)) -1.0) 0 0) (position
(+ Xg (+ Lspacer Lpreox)) -0.005 wn) "Oxide" "PolyReOxide2")

;Spacer
(isegeo:create-cuboid (position (* (+ Xg (+ Lspacer Lpreox)) -1.0) 0 0) (position
(+ Xg (+ Lspacer Lpreox)) Ypol wn) "Si3N4" "NiSpacer")
(ise:define-parameter "fillet-radius" 0.01 0.0 0.0 )
(isegeo:fillet-edges (list (car (find-edge-id (position (* (+ Xg (+ Lspacer Lpreox))
-1.0) Ypol (/ wn 2)))) (car (find-edge-id (position (+ Xg (+ Lspacer Lpreox)) Ypol
(/ wn 2)))) ) fillet-radius)

;————————————————————————————————
; DEFINING AND PLACING CONTACTS

; SUBSTRATE CONTACT
(isegeo:define-contact-set "substrate" 4.0 (color:rgb 0.0 1.0 1.0 ) "##")
(isegeo:define-3d-contact (find-face-id (position 0.01 Ysub 0.01)) "substrate")

; GATE CONTACT
(isegeo:define-contact-set "gate" 4.0 (color:rgb 0.0 0.0 1.0 ) "——")
(isegeo:define-3d-contact (find-face-id (position 0 Ypol (/ wn 2))) "gate")

; DRAIN CONTACT
(isegeo:create-cuboid (position -0.066 0 0.02) (position -0.164 -0.2 (- wn 0.02))
"Metal" "Drainmetal")
(isegeo:define-contact-set "drain_nmos" 4.0 (color:rgb 0.0 0.0 1.0 ) "##")
(isegeo:define-3d-contact (find-face-id (position -0.1 0 (/ wn 2))) "drain_nmos")
(isegeo:delete-region (find-body-id (position -0.1 -0.1 (/ wn 2))))

; SOURCE CONTACT
(isegeo:create-cuboid (position 0.066 0 0.02) (position 0.164 -0.2 (- wn 0.02))
"Metal" "Sourcemetal")
(isegeo:define-contact-set "source_nmos" 4.0 (color:rgb 0.0 0.0 1.0 ) "##")
(isegeo:define-3d-contact (find-face-id (position 0.1 0 (/ wn 2))) "source_nmos")
(isegeo:delete-region (find-body-id (position 0.1 -0.1 (/ wn 2))))

; p-WELL CONTACT (this would be connected to ground, along with the source)
(isegeo:create-cuboid (position 0.946 0 (+ subzmin 0.02)) (position 1.044 -0.2
(- subzmax 0.02)) "Metal" "pwell")
(isegeo:define-contact-set "pwell" 4.0 (color:rgb 0.0 0.0 1.0 ) "##")
(isegeo:define-3d-contact (find-face-id (position 0.98 0 0.12)) "pwell")
(isegeo:delete-region (find-body-id (position 0.98 -0.1 0.12)))
```

;————————————————————————————————
; Saving BND file
;(define SOI (part:entities (filter:type "solid?"))) (iseio:save-dfise-bnd SOI "nmos65jon.bnd")
;————————————————————————————————

; SET DOPING REGIONS AND PROFILES
; CONSTANT DOPING PROFILES
; SUBSTRATE REGION AND PROFILE
(isedr:define-constant-profile "region_1" "BoronActiveConcentration" Dop)
(isedr:define-constant-profile-region "region_1" "region_1" "region_1")

; PolySi GATE REGION AND PROFILE - Main
(isedr:define-constant-profile "region_3" "ArsenicActiveConcentration" 2e20)
(isedr:define-constant-profile-region "region_3" "region_3" "region_3")

; ANALYTICAL DOPING PROFILES
; SUBSTRATE (LATCHUP) PROFILE (IN BETWEEN THE p-WELL AND THE SUBSTRATE)
(isedr:define-refinement-window "Latchup.Profile.Region" "Rectangle" (position subxmin 1.25 subzmin) (position subxmax 1.25 subzmax))
(isedr:define-gaussian-profile "Latchup.Profile" "BoronActiveConcentration" "PeakPos" 0 "PeakVal" 5e18 "ValueAtDepth" 1e16 "Depth" 0.4 "Gauss" "Factor" 0.0001)
(isedr:define-analytical-profile-placement "Latchup.Profile.Place" "Latchup.Profile" "Latchup.Profile.Region" "Symm" "NoReplace" "Eval")

; p-WELL PROFILE OF THE NMOS DEVICE
(isedr:define-refinement-window "pwell.Profile.Region" "Rectangle" (position subxmin 0.65 subzmin) (position subxmax 0.65 subzmax))
(isedr:define-gaussian-profile "pwell.Profile" "BoronActiveConcentration" "PeakPos" 0 "PeakVal" 2e18 "ValueAtDepth" 1e17 "Depth" 0.35 "Gauss" "Factor" 0.0001)
(isedr:define-analytical-profile-placement "pwell.Profile.Place" "pwell.Profile" "pwell.Profile.Region" "Symm" "NoReplace" "Eval")

; p-WELL CONTACT PROFILE (DEGENRATE DOPING FOR p-WELL CONTACT)
(isedr:define-refinement-window "pwelltap.Profile.Region" "Rectangle" (position 0.93 0 subzmin) (position 1.06 0 subzmax))
(isedr:define-gaussian-profile "pwelltap.Profile" "BoronActiveConcentration" "PeakPos" 0 "PeakVal" 2e20 "ValueAtDepth" 1e17 "Depth" 0.06 "Gauss" "Factor" 0.0001)

(isedr:define-analytical-profile-placement "pwelltap. Profile.Place" "pwelltap.Profile" "pwelltap.Profile.Region" "Symm" "NoReplace" "Eval")

; SOURCE
(isedr:define-refinement-window "source.Profile.Region" "Rectangle" (position 0.05 0 0) (position 0.18 0 wn))
(isedr:define-gaussian-profile "source.Profile" "ArsenicActiveConcentration" "PeakPos" 0 "PeakVal" 2e20 "ValueAtDepth" 1e17 "Depth" 0.024 "Gauss" "Factor" 0.1)
(isedr:define-analytical-profile-placement "source.Profile.Place" "source.Profile" "source.Profile.Region" "Symm" "NoReplace" "Eval")

; SOURCE HALO
(isedr:define-refinement-window "HSimplant.Profile.Region" "Rectangle" (position 0.012 0.014 0) (position 0.017 0.014 wn))
(isedr:define-gaussian-profile "HSimplant.Profile" "BoronActiveConcentration" "PeakPos" 0 "PeakVal" 2e19 "ValueAtDepth" 1e16 "Depth" 0.014 "Gauss" "Factor" 0.0001)
(isedr:define-analytical-profile-placement "HSimplant.Profile.Place" "HSimplant.Profile" "HSimplant.Profile.Region" "Symm" "NoReplace" "Eval")

; DRAIN
(isedr:define-refinement-window "drain.Profile.Region" "Rectangle" (position -0.05 0 0) (position -0.18 0 wn))
(isedr:define-gaussian-profile "drain.Profile" "ArsenicActiveConcentration" "PeakPos" 0 "PeakVal" 2e20 "ValueAtDepth" 1e17 "Depth" 0.024 "Gauss" "Factor" 0.1)
(isedr:define-analytical-profile-placement "drain.Profile.Place" "drain.Profile" "drain.Profile.Region" "Symm" "NoReplace" "Eval")

; DRAIN HALO
(isedr:define-refinement-window "HDimplant.Profile.Region" "Rectangle" (position -0.012 0.014 0) (position -0.017 0.014 wn))
(isedr:define-gaussian-profile "HDimplant.Profile" "BoronActiveConcentration" "PeakPos" 0 "PeakVal" 2e19 "ValueAtDepth" 1e16 "Depth" 0.014 "Gauss" "Factor" 0.0001)
(isedr:define-analytical-profile-placement "HDimplant.Profile.Place" "HDimplant.Profile" "HDimplant.Profile.Region" "Symm" "NoReplace" "Eval")

; LDD - SOURCE
(isedr:define-refinement-window "sourceldd.Profile.Region" "Rectangle" (position 0.016 0.0 0) (position 0.08 0 wn))
(isedr:define-gaussian-profile "sourceldd.Profile" "ArsenicActiveConcentration" "PeakPos" 0 "PeakVal" 8e18 "ValueAtDepth" 1e17 "Depth" 0.014 "Gauss" "Factor" 0.1)

```
(isedr:define-analytical-profile-placement "sourceldd.Profile.Place"
"sourceldd.Profile" "sourceldd.Profile.Region" "Symm" "NoReplace" "Eval")

; LDD - DRAIN
(isedr:define-refinement-window "drainldd.Profile.Region" "Rectangle" (position
-0.016 0.0 0) (position -0.08 0 wn))
(isedr:define-gaussian-profile "drainldd.Profile" "ArsenicActiveConcentration"
"PeakPos" 0 "PeakVal" 8e18 "ValueAtDepth" 1e17 "Depth" 0.014 "Gauss" "Fac-
tor" 0.1)
(isedr:define-analytical-profile-placement "drainldd.Profile.Place"
"drainldd.Profile" "drainldd.Profile.Region" "Symm" "NoReplace" "Eval")

; Vt IMPLANT
(isedr:define-refinement-window "implant.Profile.Region" "Rectangle" (position
-0.015 0.002 0) (position 0.015 0.002 wn))
(isedr:define-gaussian-profile "implant.Profile" "BoronActiveConcentration"
"PeakPos" 0 "PeakVal" 8e18 "ValueAtDepth" 1e17 "Depth" 0.01 "Gauss" "Factor"
0.0001)
(isedr:define-analytical-profile-placement "implant.Profile.Place" "implant.Profile"
"implant.Profile.Region" "Symm" "NoReplace" "Eval")

; IMPLANT TO MITIGATE LEAKAGE (BELOW Vt IMPLANT)
(isedr:define-refinement-window "limplant.Profile.Region" "Rectangle" (position
-0.015 0.014 0) (position 0.015 0.014 wn))
(isedr:define-gaussian-profile "limplant.Profile" "BoronActiveConcentration"
"PeakPos" 0 "PeakVal" 7e18 "ValueAtDepth" 2e17 "Depth" 0.005 "Gauss" "Fac-
tor" 0.0001)
(isedr:define-analytical-profile-placement "limplant.Profile.Place"
"limplant.Profile" "limplant.Profile.Region" "Symm" "NoReplace" "Eval")

; STI Implant - Front & Back Extensions
(isedr:define-refinement-window "Window.FrontB" "Rectangle" (position -0.018 0
-0.001) (position 0.018 0.36 -0.001))
(isedr:define-refinement-window "Window.BackB" "Rectangle" (position -0.018 0
(+ wn 0.001)) (position 0.018 0.36 (+ wn 0.001)))
(isedr:define-constant-profile "Profile.ImplantB" "BoronActiveConcentration"
5e19)

(isedr:define-constant-profile-placement "Place.Implant.FrontB" "Profile.ImplantB"
"Window.FrontB")
(isedr:define-constant-profile-placement "Place.Implant.BackB" "Profile.ImplantB"
"Window.BackB")

;------------------------------------------------------------------------
-----

; DEFINE MESHING REGIONS AND MAX-MIN MESH SPACINGS
```

```
; UPPER SUBSTRATE REGION
(isedr:define-refinement-size "region_1" 0.5 0.5 0.5 0.2 0.2 0.2)
(isedr:define-refinement-window "region_1" "Cuboid" (position subxmin 0.1 subzmin) (position subxmax 2 subzmax))
(isedr:define-refinement-function "region_1" "DopingConcentration" "MaxTransDiff" 0.1)
(isedr:define-refinement-placement "region_1" "region_1" "region_1" )

; STI IMPLANT
(isedr:define-refinement-size "sti" 0.01 0.025 0.001 0.005 0.005 0.0005)
(isedr:define-refinement-window "sti" "Cuboid" (position -0.018 0 0) (position 0.018 0.36 0.002))
(isedr:define-refinement-function "sti" "DopingConcentration" "MaxTransDiff" 0.1)
(isedr:define-refinement-placement "sti" "sti" "sti" )

; STI IMPLANT-I
(isedr:define-refinement-size "sti1" 0.01 0.025 0.001 0.005 0.005 0.0005)
(isedr:define-refinement-window "sti1" "Cuboid" (position -0.018 0 0.998) (position 0.018 0.36 1))
(isedr:define-refinement-function "sti1" "DopingConcentration" "MaxTransDiff" 0.1)
(isedr:define-refinement-placement "sti1" "sti1" "sti1" )

; LOWER SUBSTRATE REGION
(isedr:define-refinement-size "region_12" 0.75 0.75 0.75 0.5 0.5 0.5)
(isedr:define-refinement-window "region_12" "Cuboid" (position subxmin 2 subzmin) (position subxmax Ysub subzmax))
(isedr:define-refinement-function "region_12" "DopingConcentration" "MaxTransDiff" 0.1)
(isedr:define-refinement-placement "region_12" "region_12" "region_12" )

; CHANNEL REGION
;(isedr:define-refinement-size "R.Channel" 0.01 0.01 0.1 0.002 0.002 0.1)
(isedr:define-refinement-size "R.Channel" 0.005 0.005 0.05 0.002 0.002 0.1)
(isedr:define-refinement-window "R.Channel" "Cuboid" (position (* Xg -1.0) 0 0) (position Xg 0.05 wn))
(isedr:define-refinement-function "R.Channel" "DopingConcentration" "MaxTransDiff" 0.1)
(isedr:define-refinement-placement "R.Channel" "R.Channel" "R.Channel" )

; SOURCE/DRAIN REGION
(isedr:define-refinement-size "sourcedrain" 0.02 0.02 0.1 0.02 0.02 0.1)
(isedr:define-refinement-window "sourcedrain" "Cuboid" (position -0.18 0 0) (position 0.18 0.1 wn))
```

(isedr:define-refinement-function "sourcedrain" "DopingConcentration" "MaxTransDiff" 0.1)
(isedr:define-refinement-placement "sourcedrain" "sourcedrain" "sourcedrain")

; Vt & LEAKAGE IMPLANT REGIONS
(isedr:define-refinement-size "implant" 0.01 0.01 0.1 0.002 0.002 0.1)
(isedr:define-refinement-window "implant" "Cuboid" (position -0.018 0 0) (position 0.018 0.07 wn))
(isedr:define-refinement-function "implant" "DopingConcentration" "MaxTransDiff" 0.1)
(isedr:define-refinement-placement "implant" "implant" "implant")
; p-WELL CONTACT REGION
(isedr:define-refinement-size "ptap" 0.1 0.1 0.2 0.05 0.05 0.2)
(isedr:define-refinement-window "ptap" "Cuboid" (position 0.93 0 subzmin) (position 1.06 0.1 subzmax))
(isedr:define-refinement-function "ptap" "DopingConcentration" "MaxTransDiff" 0.1)
(isedr:define-refinement-placement "ptap" "ptap" "ptap")

; p-WELL CONTACT REGION-I
(isedr:define-refinement-size "ptap1" 0.02 0.02 0.1 0.005 0.005 0.05)
(isedr:define-refinement-window "ptap1" "Cuboid" (position 0.93 0 0) (position 1.06 0.1 wn))
(isedr:define-refinement-function "ptap1" "DopingConcentration" "MaxTransDiff" 0.1) (isedr:define-refinement-placement "ptap1" "ptap1" "ptap1")

; ION TRACK
(isedr:define-refinement-size "itrack" 0.01 0.5 0.01 0.005 0.5 0.005)
(isedr:define-refinement-window "itrack" "Cuboid" (position -0.07 0 (- (/ wn 2) 0.06)) (position -0.15 Ysub (+ (/ wn 2) 0.06)))
(isedr:define-refinement-function "itrack" "DopingConcentration" "MaxTransDiff" 0.1)
(isedr:define-refinement-placement "itrack" "itrack" "itrack")

;————————————————————————
; Save CMD file
(sdedr:write-cmd-file "nmos_msh.cmd")

;————————————————————————
; Meshing structure
(ise:build-mesh "mesh" "-P" "nmos_msh")

The above code generates the 3D NMOS device shown in Fig. 5.2. This code generates two files (nmos_msh.grd and nmos_msh.dat) for the 3D NMOS transistor. Both these files are used with Sentaurus-DEVICE for mixed-level simulations.

A.1.1 *Code for Mixed-Level Simulation of a Radiation Particle Strike Using Sentaurus-DEVICE*

To simulate a radiation particle strike at the drain of the NMOS transistor of the INV shown in Fig. 5.1, the following code was used with Sentaurus-DEVICE simulator.

```
# define the n-channel MOSFET;
Device NMOS {
Electrode {
{ Name="source_nmos" Voltage=0 }
{ Name="drain_nmos" Voltage=0 }
{ Name="gate" Voltage=0}
{ Name="pwell" Voltage=0 }
{ Name="substrate" Voltage=0}

}
# Define input and output files for simulation
File {
Grid = "nmos_msh.grd" #NMOS transistor file
Doping = "nmos_msh.dat" #NMOS transistor file
Plot = "nmos"
Current = "nmos"
Param = "mos"
}

# Physical models to be applied in the simulation
Physics {
Mobility( PhuMob ( Arsenic ) HighFieldsat Enormal )
Fermi
EffectiveIntrinsicDensity( OldSlotboom )
Recombination ( SRH Auger )
Hydrodynamic( eTemperature )
#Heavy ion strike
HeavyIon (
PicoCoulomb
Direction=(0,1,0)
Location=(-0.11,0,0.5)
Length=2
Time=1e-9
LET_f=0.1 #0.1pC corresponds to 10MeV-cm2/mg
wt_hi=0.03
Gaussian
)

}
}
```

```
File{
Plot = "nmosparticlestrike_n@node@.dat"
SPICEPath = "."
Current = "nmoslet10mm_n@node@.plt"
}

#Define the electric circuit which is to be simulated
System{
NMOS nmos("source_nmos" = 0 "pwell"=0 "gate"=n1 "drain_nmos" = n2 "sub-
strate"=0)
pmos m0 (n2 n1 n3 n3) {w=4e-6 l=0.065e-6 as=0.52e-12 ad=0.52e-12 ps=4.26e-6
pd=4.26e-6}
Vsource_pset v1(n3 0){pwl = (0 0 100p @vdd@ 10e-9 @vdd@)}
Vsource_pset vin(n1 0){pwl = (0 0 150e-12 0 5e-9 0)}
Capacitor_pset c1 (n2 0) {capacitance=25e-15}
Plot "nmosstrike_n@node@.plt" (time() n2 i(nmos n2) i(m0 n2) i(c1 n2))
}

#Specify solution variables to be saved in the output plot files
Plot {
eDensity hDensity eCurrent hCurrent
equasiFermi hquasiFermi
eTemperature
ElectricField eEparallel hEparallel
Potential SpaceCharge
SRHRecombination Auger AvalancheGeneration
eMobility hMobility eVelocity hVelocity
Doping DonorConcentration AcceptorConcentration
ConductionBandEnergy ValenceBandEnergy
HeavyIonChargeDensity
}

#Define few settings for the numeric solver
Math { Extrapolate
Derivatives
Newdiscretization
RecBoxIntegr
Method=ILS
RelErrControl
Iterations=20
notdamped=100
Number_of_Threads = 4
Wallclock
}
```

```
#Define a sequence of solutions to be obtained by the solver
Solve {
Coupled (Iterations=100) {Circuit}
Coupled (Iterations=100) {Poisson}
Coupled (Iterations=100) {Poisson Circuit}
Coupled (Iterations=100) {Poisson Contact Circuit}
Coupled (Iterations=100) {Poisson Hole Contact Circuit}
Coupled (Iterations=100) {Poisson Hole Electron Contact Circuit}


NewCurrentFile="transient_n@node@"
#Define transient simulation parameters
Transient (
InitialTime=0 FinalTime=0.99e-9 InitialStep=1e-12 MaxStep=1e-10
Increment=1.3)
{
Coupled {nmos.poisson nmos.electron nmos.hole nmos.contact circuit }
}
Transient ( # Take very small time step during the heavy ion strike
InitialTime=0.99e-9 FinalTime=1.1e-9 InitialStep=1e-13 MaxStep=1e-12
Increment=1.3)
{
Coupled {nmos.poisson nmos.electron nmos.hole nmos.contact circuit }
Plot (FilePrefix="invconstdmm_n@node@_10" Time=(1.0e-9; 1.01e-9; 1.02e-9;
1.035e-9; 1.05e-9;
1.07e-9; 1.09e-9) NoOverwrite)
}
Transient (
InitialTime=1.1e-9 FinalTime=4e-9 InitialStep=1e-12 MaxStep=1e-10
Increment=1.3)
{
Coupled {nmos.poisson nmos.electron nmos.hole nmos.contact circuit }
Plot (FilePrefix="invconstdmm_n@node@_10_1" Time=(1.11e-9; 1.13e-9; 1.15e-
9) NoOverwrite)
}

}
```

Index

T

V

FSC
www.fsc.org
MIX
Papier | Fördert
gute Waldnutzung
FSC® C083411

Kolibri 360 GmbH
Ferdinand-Jühlke-Straße 7
99095 Erfurt, Deutschland
produktsicherheit@kolibri360.de